Saving Molly

A Research Veterinarian's Choices

■ ■ ■

DR. JAMES MAHONEY

ALGONQUIN BOOKS
OF CHAPEL HILL
1998

Published by
ALGONQUIN BOOKS OF CHAPEL HILL
Post Office Box 2225
Chapel Hill, North Carolina 27515-2225

a division of
Workman Publishing
708 Broadway
New York, New York 10003

Printed in the United States of America.
Published simultaneously in Canada by Thomas Allen & Son Limited.
Design by Anne Winslow.

Library of Congress Cataloging-in-Publication Data
Mahoney, James, 1940–
Saving Molly : a research veterinarian's choices / James Mahoney.
p. cm.
ISBN 1-56512-173-2
1. Mahoney, James, 1940– . 2. Veterinarians—United States—Biography. 3. Dogs—United States—Biography. 4. Human-animal relationships. 5. Vivisection—United States. 6. Laboratory animals—United States—Anecdotes. I. Title.
SF613.M3A3 1998
636.089'092—dc21 98-11596
[B] CIP

10 9 8 7 6 5 4 3 2 1
First Edition

To my wife, Marie-Paule, longtime friend and honest mirror, and to my daughter, Nathalie, and the boys, Pádraig and Christopher—a family that always stood by me and the *crathers.*

The universe resounds with the joyful cry "I am!"

—SCRIABIN

Introduction

By Roger A. Caras

■ ■ ■

JIM MAHONEY, VETERINARIAN and philosopher, is, in a very real sense, a man trapped in the middle of a storm. He is caught in the center of the rancorous controversy over animal research.

That controversy rages not only around him but inside him as well. Jim Mahoney is a man divided. First, of course, there is the Jim Mahoney who can tell us the story of how he saved Molly, the dog whose struggle to overcome all of her various illnesses is compelling and beautiful. There is that Jim Mahoney. Molly's Jim.

There is also Dr. Mahoney of LEMSIP (the Laboratory for Experimental Medicine and Surgery in Primates). For years now, he has been fighting to make things better for creatures whom he considers close relatives. Delivering better care than perhaps any other research veterinarian in the world, he is the chimpanzee's doctor. He walks among them, talks to them, and obviously loves them. Still, the chimps he cares for so tenderly and with such passionate understanding are chimps used in laboratory protocols, as a means of testing hepatitis vaccines as well as new drugs being developed in the war on AIDS. Dr. Mahoney partakes in a complex of activity known collectively—and frequently hatefully—as vivisection or biomedical research.

The very term, vivisection, arouses strong emotional reactions on both sides. On one side stand most medical scientists, rigid and unyielding; on the other, a host of philosophers and ethicists. Science claims that it knows no alternative to using animals in its urgent search for cures for AIDS, cancer, and birth defects, while those most adamantly opposed to research involving animals claim that scientists have the alternatives at their disposal but for a variety of reasons refuse to use them. Those most furiously engaged in this battle say it doesn't matter if we know of alternatives to using animals or not—we simply do not have the right to *use* animals for our own good, or for any other purpose. These activists swear that they will never deviate from that position, and one is inclined to believe them. What we have, then, is rigidity squared off with rigidity.

In the middle stand Molly's Jim as well as Dr. Mahoney, the chimpanzees' doctor, both packaged inside one gentle, caring man trying to calm the emotions, soften the rhetoric, and most important, make sense of it all. There is conflict within him that one doubts can be resolved. Jim Mahoney knows only too well that his patients are only once removed from him—and us—by 1 or 2 percent of their genetic material, that we are well within 2 percent of being chimpanzees ourselves. And so, Jim Mahoney seems a man destined to remain suspended. He is conscientious and innovative, tirelessly working to improve the lives of his chimps as infants and adults. But he can never forget that in the end they are to be used as models for human beings in tests that, although not painful, require the chimps to be held in relatively small cages, generally in isolation. For animals as social as the chimpanzee, any form of isolation is painful, and no one knows this better than Dr. James Mahoney.

And then there is Molly. She would have died in Jamaica if it hadn't been for Jim, but he brought her home. He cared for her lovingly and watched her grow into a remarkable little creature I have had the pleasure of knowing (and who, I might add, uses her paws as no other dogs do, or at least not any other dog I have known).

Who, then, is the real Dr. James Mahoney—Molly's Jim only, or the rider of the storm who happens to own a dog about whom he is sentimental? He is, I would contend, both men. Like so many other people who have become immersed in the bewilderingly complex world of the human/animal relationship, he is caught up in a myriad of ambiguities. Dr. Mahoney's skills as a scientist and veterinarian are finite, but less finite is our collective guilt and, in particular, that felt and suffered by Molly's Jim.

Acknowledgments

■ ■ ■

I AM DEEPLY INDEBTED to everyone at Algonquin, but especially to Memsy Price and Dana Stamey, who had the patience to work with me and not give up.

I also thank all the technicians and other staff at LEMSIP who, over the years, gave so much love. I have mentioned only a few of them by name, yet I could relate at least one story about every single one of them—some touching moment, some special deed, some comical event. But I admired them most of all for being a team able to pull together when things got tough. I saw this every time there was a medical emergency with one of the chimps. When I would hear the announcement over the PA system, I would immediately drop what I was doing and rush to the animal's room. Even before I arrived on the scene, the technicians would have already paired off and formed teams; one pair would be administering cardiopulmonary resuscitation, another pair would have inserted an endotracheal tube into the animal's windpipe, and yet another would have set up an IV line to administer adrenaline or some other stimulant into the blood stream, while the electrocardiograph bleeped out its wiggly trace in the background. I would stand on the sidelines, like a maestro ignored by the orchestra, hardly needing to lay a hand on the patient. The technicians didn't always succeed, and sometimes, the animal died despite their valiant efforts. When they failed, the technicians at least had one another's comfort and the knowledge that they had done all that was humanly possible to save the animal. But when they were successful, they experienced a high I doubt few other people ever reach in the normal course of their daily work.

Saving Molly

Prologue

■ ■ ■

THIS IS THE STORY of a brave little dog by the name of Molly—a bush dog, blind in one eye, and poor-sighted in the other. Born under the crawl space of a tiny house in southwest Jamaica, Molly was the runt of the litter, barely half the size of her brothers and sisters. At the time I met her, she suffered life-threatening anemia, brought on by a massive flea infestation. She was riddled with intestinal parasites and weakened by uncontrollable diarrhea. The lids of both her eyes were matted tightly shut with pus, and heaven knows what other medical problems afflicted her.

I am a veterinarian, and my wife, Marie-Paule, and I came across the little puppy while we were on vacation. She was just three and a half weeks old. We took her under our wing for a while, hoping that by the end of our brief stay, she would be well on the road to recovery. But things didn't turn out that way.

As I cared for Molly over the subsequent months, struggling to help her overcome one serious health problem after another, I began to question my own life and its meaning. I remembered other special animals I had known, who, for one reason or another, also had to struggle. Sometimes their problems were health related, like Molly's; sometimes they had to do with how we as human beings use animals in our various pursuits and needs; and sometimes they occurred simply because of ugly human politics.

I take care of, and experiment on, monkeys and chimpanzees at LEMSIP (the Laboratory for Experimental Medicine and Surgery in Primates at New York University) in the difficult, and often frustrating, search for vaccines and cures against major human diseases such as hepatitis and AIDS. It is serious work, and I see no alternative to using animals in research if we are to continue making the breakthroughs in human health that have occurred in the last fifty years or more. But the animals sometimes pay a high price for their involuntary contribution, not always physically, as many might think, but more emotionally and psychologically. Caring for these animals forces me to confront a dilemma which increasingly haunts me: How can someone like me, a veterinarian trained to provide compassionate care for animals, devote his career to what many see as the cruel and inhumane world of animal research?

When, more than thirty years ago, I started out as a newly qualified veterinarian attending the little farms of the Southern Uplands of Scotland, I thought life was simple. All I wanted was to be a good horse doctor; I knew where I was going and understood the purpose of what I was doing. I had wanted to be a veterinarian since I was seven years old. I pictured myself in a neat white laboratory coat with a stethoscope around my neck as I tried valiantly to save some little dog or cat from distemper or automobile injuries. I even visualized myself as a horse doctor in Ireland, attending Thoroughbreds on the open plain of the Curragh of County Kildare or in the gentle green vales of Tipperary.

As a child, I loathed the idea of research, without understanding in the least what it entailed. The antivivisection posters in the London Underground stations are still vivid in my memory. The mainly black-and-white drawing depicted a dog, a cat, and a rabbit huddled together for protection while a masked researcher in the background wielded a scalpel, its blade dripping with bright red blood.

Horror stories of the most terrible cruelties to research animals abounded in the newspapers: dogs kept in wire cages, in the filthiest of conditions in poorly ventilated back rooms of teaching hospitals, with barely enough room to stand up; cats crowded into similarly cramped cages, reeking of urine, with inadequate food and stale water to drink. All this in a country that had always prided itself on its compassionate attitude toward animals.

Images of such barbarities fueled my desire to go forth in shining armor, sword ready to slash and flail. After all, I wanted to be a vet because I loved animals. "Don't we all!" scoffed one of the professors interviewing me at the Royal College of Veterinary Surgeons in London. He was sitting behind a vast oak table that separated him and the other examining professors from me, sitting lonely on my chair. His colleagues guffawed in agreement. Then I had to admit to them that, no, I had never witnessed the birth of a foal, or a calf, or even a puppy. "Well, you must have at least seen a chicken lay an egg," one of the professors said. The world seemed to stop spinning at that moment, and I knew I had failed the interview. Still, I was determined to be a veterinarian.

I spent a year working on a farm, where I did see foalings, calvings, and whelpings, although to this day I have yet to see a hen lay an egg. I later made it into veterinary school at Glasgow University—the same school that had graduated James Herriot more than thirty years before—but only by the skin of my teeth and probably because I lied to the professors on the examining board. On a friend's advice, I disingenuously told the board I had an interest in research. That one day I would actually find myself in research, and with monkeys and chimpanzees as my patients, was the furthest thought from my mind.

I have never seen myself as a spokesman for animal research. My mission, as I see it, is to encourage a gentler, more compassionate approach toward animals in the laboratory. A few years ago while I was attending a scientific conference, a fellow researcher said to me in the kindest way that she thought I represented the very epitome of the humane scientist. It occurred to me that there is nothing remarkable about being humane; it's being *inhumane* that's remarkable. You don't pat yourself on the back at the end of your life and say, "I've been a good person in my life; I never murdered anyone."

The real issue is compassion, which can only be reached through empathy. I have tried to see the frightening and unnatural world of the research laboratory through the eyes of the animals confined in their cages, and I've tried to act accordingly. This meant remembering that to a rhesus monkey or a baboon, it's a threat to stare them in the eye. A great, brawny chimpanzee, or a tiny marmoset, on the other hand, demand fixed attention from you, lots of eye contact. "You don't blow into a room like a whirlwind with syringe or dart gun in hand to anesthetize an animal, and then turn to disappear just as rapidly," I would tell new lab technicians. "You must stay with the animal and talk him through what must undoubtedly be a frightening sensation, as the world begins to spin before his eyes and he slips into unconsciousness."

I do not, of course, carry out my work alone. I'm assisted by a group of the greatest technicians imaginable. *Technicians* is a word that might conjure up images of cold, calculating characters going about their work in an efficient and clinical fashion, devoid of feeling for the animals in their charge. Nothing could be further from the truth. The old term was *caretaker,* or worse still, *animal handler.* The more modern term applied to these people is *caregiver,* a term that may have been coined by Jane Goodall herself, the very essence of gentleness and compassion. But the people at LEMSIP who look after the primates are much more than caregivers. They are, by their own definition, technicians —highly trained medical professionals, justly proud of their abilities and accomplishments.

Most of the technicians at LEMSIP are not comfortable using animals in research, but, concluding that it will be done by someone or other in some laboratory or other no matter what, they elect to be part of the process because at least then they have the satisfaction of knowing it is done well, that the animals' needs are being carefully considered.

And now LEMSIP faces a crisis—not the first in its tumultuous thirty years of existence, but certainly its most serious. Forced to retire, laid off, fired—the explanation varying with whomever you speak—there is no Dr. Jan Moor-Jankowski, founder and director, to rescue the lab as he did in 1980 when it came within ten days of closing. I have been given the unenviable distinction of becoming the acting director, the captain of the sinking ship.

Despite the lab's contributions to human medicine, New York University wishes to divest itself of LEMSIP.

Faced with its own massive financial problems brought on largely by new managed health care systems, the university's medical center no longer sees LEMSIP's mission as consistent with its own. LEMSIP has become a pimple on the bottom of the elephant, small but very aggravating. The Medical Center has entered into an agreement with the Coulston Foundation, a New Mexico-based research laboratory which specializes in toxicology and has the largest chimpanzee colony in the world, to take over LEMSIP after a six-month transition period. Whether the Coulston Foundation takes over or not, it will get one hundred of LEMSIP's approximately 240 chimps, and perhaps a great deal more unless I am able to find retirement homes for the remainder.

On top of the sadness and tragedy of the proud little laboratory's disappearance, there is bitterness and rancor. Many of the animal rights groups distrust me, seeing me as selling out the animals; many in my own university see me as a renegade. By the time this book reaches publication, most, and perhaps all, of LEMSIP's staff—some with as many as fifteen, twenty, or even twenty-five years of faithful service—will have lost their jobs.

But away from the realities of the research laboratory and the often bitter politics of modern medicine, as I sweated through the Jamaican nights with a dog of modest origins, I remembered something about myself. Saving Molly would be an obvious act for me; there would be no debate over costs or ultimate benefits. My experience with Molly began to crystallize thirty years of rumination about my life with animals. This, then, is also the story of my reawakening and the animals that were there for it.

Chapter One

■ ■ ■

I WALKED OUT ONTO THE edge of the low cliff that overlooked Calabash Bay. Another glorious Jamaican day was about to begin. What a welcome change, if only for seven or eight days, to get away from the constant demands of my job.

Marie-Paule and I can usually spare only a week or so for a holiday, mainly because of her busy schedule teaching French at a private school and immersion language courses at the state university. Moreover, I often have to schedule my vacation months in advance, so that there's no clash with a particularly demanding research project; I also can't leave if Mike and Dave, my two breeding technicians, and I anticipate a problem with one of the chimps giving birth, and we might have to perform a cesarean section, induce labor, or provide special care to the newborn infant. So every moment counts for us. Over the years we've come to find our Jamaica vacation truly begins with the first, and always unexpected, meeting of the goats and brown cows that graze along the roadside as we leave the outskirts of Montego Bay. The tensions of daily life seem to melt away as we decelerate into the calm rhythm of Jamaica.

The sea that early Sunday morning was streaked turquoise and aquamarine. Not a single whitecap was in sight, and the gentle breeze of the trade winds held the rising heat at bay. I scanned the fine sandy beach from one end to the other. About half a mile away I could see early rising fishermen cleaning their brightly painted wooden long boats beached high up on the sand and moored to rocks and twisted old trees. The breeze carried the giggles and screeches of little children already at play around the collection of tin-roofed houses that make up the small fishing village, and the smell of wood fires lit to cook breakfast wafted toward me.

A pair of magnificent frigate birds, which seem to be leftovers from prehistoric times, were already patrolling high in the sky, some distance out to sea, their motionless crescentic wings outstretched like pterodactyl pennons. Two sleepy-looking turkey vultures, or John crows as the islanders call them, were perched on a limb of a scraggly tree. The birds sat waiting for the first thermals, which would carry them effortlessly aloft. Their wings, like the black cloaks of evil witches, were outstretched, the undersides facing up to absorb the heat of the early morning sun and bring the birds' heart rates back up to normal after their overnight torpor.

Halfway to the gathering of fishermen I noticed a woman kneeling in the sand, her back to the water. She appeared to be washing something in the sea, and every now and again round, ball-like objects pinged away from her hands. What on earth is she doing? I wondered. I peered at her through my bird-watching binoculars and saw that the woman's long silver hair was tied in a loose ponytail that hung forward over one shoulder. She was obviously European. As I strained to see through the glasses, I realized that she had a bundle of newborn puppies scooped up in the skirt of her long white dress. She was holding one puppy at a time, washing it with her now saturated skirt, and then releasing it to run away, yapping and barking with excitement.

I have always been moved by the sight of newborn puppies and decided I should go down and investigate more closely. I walked to the edge of the land that surrounded the little villa where Marie-Paule and I were staying and climbed over the coral-stone wall, amid the wild profusion of red hibiscus and purple bougainvillea, which danced giddily in the breeze, scraping the inside of my knee as I went. From there it was just a short walk down a steep flight of wooden steps to the beach.

The woman was still there, kneeling in the surf, one last puppy to be washed. She was handsome, with wind-burnt skin, a lithe body, and delicate hands. I wasn't sure how old she might be, but, as the French would demurely say, she was of "*un certain âge*."

"Good morning," I said. "What are you doing with the puppies?"

"I'm washing them, can't you see?" she replied. "The seawater helps get rid of any fleas they might have."

To my immediate surprise, I realized that she had the unmistakable lilt of the Jamaican accent, or what you might call an educated version of it, somewhat reminiscent of County Kerry, in the southwest of Ireland, or the Outer Hebrides, the islands off the northwest coast of Scotland. It's little wonder that Jamaicans have a sort of Irish-cum-Scots cadence, for, to the Gael's eternal shame, many of the overseers on the slave plantations, up until the early nineteenth century, came from Ireland and Scotland. It's not uncommon to find Jamaicans with names like Murphy and Flynn, or Sinclair and Forbes, especially around Treasure Beach.

I dallied a while, chatting with the lady and playing with the puppies. There were seven of them, three and a half weeks old. Four were black, with a few streaks of brown around the cheeks and paws. The other three were white and had odd patches of black or brown on one ear or at the side of the head or base of the tail. As I scooped them up, one at a time or in squirming bundles, to give them a cuddle, I realized they still had that wonderfully musky, dank odor, characteristic of newborn pups, like the smell of wet straw.

"How do you like our island?" the lady asked as she squinted her eyes against the sun to look up at me. This is almost invariably the first question that Jamaican's pose to the visitor. "I love it," I replied, explaining with pride how my wife and I had been coming to Jamaica once or twice a year for the past dozen years or more, first with our three children, Pádraig, Nathalie, and Christopher, then by ourselves when the children had grown up. We now looked upon the island as our second home.

It was not only the climate that enticed us, but also the wild beauty of the island and its quaint little villages with names like "Wait-A-Bit," "Quick Step," "Maggotty," and "Barbecue Bottom," as well as the vast, desolate regions of the Cockpit Country and the Dry Harbour Mountains, which I love to explore.

Most of all, it was the families we had got to know over the years that brought us back time and again, and the children we had seen grow up to have children of their own. We shared in their happiness, and sometimes in their grief. We saw the proud smile of Ina, as she told us about her son,

Ricky, a member of the Jamaican bobsled team that had competed in the 1992 Winter Olympics, captivating sports fans worldwide. We also saw the tears roll down Norma's cheeks as she recounted how Hurricane Gilbert had destroyed her little house. Each visit was a kind of homecoming.

I continued to watch the puppies on the beach for a while. They were now roughhousing with one another, yellow sand stuck to their wet noses and muzzles, yapping and barking with delight. "Where is their mother?" I asked.

"Oh, Molly's up at the house, taking a little time off from her motherly duties," the woman replied, nodding toward the little house that lay back above the low cliff. So, this was our neighbor, I realized. "May I come by and see her later?" I asked. "I would love to take some photographs of her and the puppies."

"You would be most welcome," she replied. "Perhaps you and your wife would come to tea, in the afternoon."

After thanking the lady for her kind invitation, I climbed back up the rugged path to the villa, scraping the skin off the inside of my knees for the second time as I scaled the sharp coral wall. Marie-Paule had just returned from her early morning walk. "You won't believe what I just saw," I said, and I told her about the puppies and the lady on the beach. "The lady's Jamaican," I added. "Yes," Marie-Paule replied, somewhat impatiently. "Remember, I told you about them last night." I suppose I'd been too tired after the previous day's long journey to pay attention to what she had said.

"She's a very nice lady," Marie-Paule continued. "Her name is Miss June, and she was born and raised in Jamaica."

Marie-Paule and I spent the rest of the day unwinding, swimming, and relaxing with our books. Around three o'clock in the afternoon, after changing out of our bathing suits into something more respectable, we made our way to Miss June's little house a short distance down the winding, narrow road. I had my camera and a good supply of film, eager to capture as many snapshots of the puppies as I could. Little did we know what tea with Miss June would bring, and how our short, restful vacation would be turned upside down.

Chapter Two

■ ■ ■

Miss June's house was a quaint one-story, two-room wooden affair with white-painted walls and blue trim, a white-painted corrugated tin roof, and a verandah with a floor of rich, highly polished wood. The house was set back from the road behind a low stone wall, hidden amongst a profusion of flowering bushes and trees.

As Marie-Paule and I walked in through the open gate, we had to duck every now and then to avoid the hanging branches of overgrown bushes. Half a dozen clucking hens, some with yellow, golf-ball-sized chicks following closely behind, scurried amongst the bushes, stopping frequently to peck the ground in search of food. A large white cockerel, obviously alarmed by our intrusion, stood erect, his brilliant red wattles and head comb flapping violently with each threatening flick of his head and coarse "crrr" of his challenging call.

I shouted out a greeting so as not to frighten Miss June

by our sudden appearance. We found her sitting outside at a low wooden table on the grass, together with three tiny children and a heavyset black woman, whom she introduced as Miss Maisy. Miss June ran a nursery school in the village, we soon discovered, and she and Miss Maisy were teaching Bible class to the three children.

As we were going through the ritual of greeting Miss June and Miss Maisy, introducing ourselves and then apologizing for being a nuisance, the puppies suddenly scampered out from under the elevated foundation of the house and ran over to sniff and nip my ankles. I crouched down to play with them and they clambered over one another to gnaw on my fingers and wrists. Looking up at Miss June, I asked, "Where is their mother?"

"Oh, Molly's off visiting Jake's Restaurant," she replied. "The owner is very fond of her, and he prepares her a little something special to eat each evening before everyone comes for dinner." I had noticed earlier in the day the blue-painted wooden sign in the lane outside the gate to Miss June's. Large white lettering and an arrow pointed out Jake's Restaurant, about half a mile farther on down the narrow, meandering road. I remembered seeing the sign outside the restaurant, announcing, in all the colors of the rainbow, SWIMMING, BEACH BAR (OPENING SOON), LUNCH-DINNER, and finally, DANCING. Jake's is described in one of my travel guides as "the spiffiest place" in Treasure Beach. So Molly is a gourmet, I thought to myself, and probably likes a touch of the high life, too.

With that, Molly suddenly appeared, not from the direction of the lane, but from the beach. This route from Jake's must have saved her a quarter of a mile's traveling, a very important consideration for a dog in a hurry. She ran to greet me, tail wagging, bottom wiggling, long pink tongue hanging out the side of her mouth, pendulous breasts and elongated nipples swaying from side to side with every movement of her body. Molly was a bush dog if ever I saw one, the same sort of dog I had seen in West Africa roaming the countryside and scavenging around the villages. Slight of build, quick and agile in movement, with an open, friendly expression on her face, Molly was just the way God had intended dogs to be, without any distortion by man's interference. She was basically white all over, with black ears, a broad black patch that covered the left side of her head, and a splattering of blue-black spots over her muzzle, shoulders, forelegs, and underbelly.

Bush dogs, more properly called pariah dogs—a term that I very much dislike because of its demeaning connotation—are so thoroughly mixed, in a genetic sense, that I think they should be regarded as a species unto themselves, distinct from the domestic dog known scientifically as *Canis familiaris*. After the wolf, coyote, and dingo, they are probably the closest true representatives of the wild dog family. Unlike their highly inbred relatives, the domestic dogs of pet owners the world over, bush dogs don't suffer from the myriad hereditary diseases and malformations common to most "pure" breeds. Their eye lenses don't drop as they do in toy poodles; their hips don't become dislocated as they do in retrievers; and their kneecaps don't pop off sideways as they do in Alsatians. Medium in size, bush dogs are usually a pale biscuit brown color all over, although they do come in other shades, including white (like Molly), tan, and even black, with a mixture of other hues thrown in for good measure. Yet they all have certain physical features that distinguish them from their domestic cousins. There is a strong element of greyhound in them: the sleek body, deep chest from spine to sternum, but narrow from side to side, thin, wiry legs, long face, and ears erect or able to be erected at a moment's notice.

You shouldn't usually fear being bitten by a Jamaican dog. They are placid and their success in staying alive is undoubtedly due to the fact that they have reached an understanding with their human counterparts, child and adult alike. They are incredibly intelligent, and know not to chase people's goats or chickens. All of these dogs belong more or less to someone, yet, because they are mostly free to roam, they are, in a sense, truly wild. I have no doubt that Molly and most of the other dogs on the island are descended from the dogs that must have accompanied the African slaves to Jamaica in the eighteenth and early nineteenth centuries, because they have so many similarities with the bush dogs of West Africa.

It so happened that, within three days after the end of Marie-Paule's and my vacation in Jamaica, that mid-August in 1993, I would set off on a previously planned trip to the Gambia, a tiny sliver of a country on the west coast of Africa. There I would meet a bush dog who reminded me very much of the Jamaican Molly. I was visiting a chimpanzee rehabilitation project located on a group of riverine islands about 175 miles upstream in the River Gambia National Forest. With the punctuality of an alarm clock, Munya, as the bush dog was called, would awaken me at seven o'clock every morning by his noisy entrance into the center of the research camp. The sound of his feet thumping against the hard-packed earth echoed through the still predawn air. He was dropping in for his daily visit from his village, situated about three miles away through the bush. Coming to from deep sleep, in my mosquito-screened cot, I would wonder each morning how such a lightweight dog could make so much noise. He was only an inch or so taller at the shoulder than the Jamaican Molly, and must have weighed no more than thirty-five pounds. His body had the same sleek, almost greyhound-like shape and configuration as Molly's, but, unlike her, he was a pale biscuit-brown color all over, except for a long splash of white down his chest. He also had the same intelligent, open face. The points of his ears had been cut off when he was a puppy, a common practice in Africa, to avoid dogs' having their ears torn in the constant territorial battles with one another. I became very attached to Munya, and we spent a great deal of time together every day, until he would return to his village each evening before sunset.

One day, he and I went for a long walk together through the bush and I discovered the reason for Munya's noisy early-morning behavior. We followed a narrow, ill-defined track through the dense foliage created by the constant travels of people passing back and forth between the research camp and Munya's village. On the trail, Munya walked close by me, his head pressed tightly against the outside of my knee, his stride in exact keeping with my gait. His unusual behavior made it quite difficult for me to walk and became all the more pronounced as we proceeded. Every time we approached a bend in the track, Munya raced forward, came to a dead standstill, and craned his neck around the corner to peer intently through the overhanging branches at the ground ahead. What on earth is he doing? I wondered. Then it dawned on me. He was making sure there were no snakes or other creepy-crawlies that might be lying in wait for the unsuspecting traveler. Having determined that the coast was clear, Munya would look back at me and wag his tail, as if to say "You can come along, now; everything's all right." He would wait for me to catch up to him before continuing on the journey, his head again pressed against my knee. I realized that his clangorous stride into camp each morning served a similar purpose—to frighten off any snakes that might be lying across his path.

Even after a lifetime of being around dogs and working as a vet, I realized, once I started to write this book, that I knew relatively little about the dog's history, and certainly nothing about bush dogs. As a child, I imagined early primitive people sitting around their wood fires at night, surrounded by the threatening darkness, and how, as they cooked their bison meat, the appetizing aromas attracted wild dogs from near and far that approached cautiously, bellies dragging the ground, ears drawn back in submission. Perhaps one man, out of fear that the dogs might attack or because he felt a sympathy for their hunger, might have cast one or two scraps out into the darkness. Little by little, over time, this ritual led the man to throw the food less far into the night, and the dogs crept ever closer to the fire, until a relationship of mutual and lasting trust developed. But this was a romantic view, I grew to realize.

If any one could set me straight on the history of dogs, I thought, it would be Roger Caras. He had authored several books on dogs, and his most recent publication, *A Perfect Harmony,* describes the intertwining lives of animals and human beings throughout history. I had got to know Roger when he made a couple of visits to LEMSIP during the 1980s to film the chimps and monkeys for science programs when he was Special Correspondent for Animals and the Environment on ABC Television. He had also shown me around the new headquarters of the American Society for Prevention of Cruelty to Animals in upper Manhattan when he became president.

I put a call through to Roger at the ASPCA, and he agreed to meet me for dinner at a Manhattan restaurant to discuss dogs. His theory of how dogs became domesticated is quite different from my childhood idea. He maintains that early man, somewhere between ten and fourteen thousand years ago, was forced to come up with a new idea for procuring food, especially during deepest winter when hunting was particularly difficult. He would locate a wolf's lair, usually a cave, scare off the mother with flaming torches, and steal the cubs as a live, fairly easily managed source of future meat. But, every now and again, there would be one wolf pup that was different from the rest of the litter—a little more affectionate perhaps, with more of a tendency to lick the hand of the feeder, rather than snap at it. No one would have the heart to kill this pup, and it would be treated as a pet. Over time, the process would be repeated, and as such wolf cubs grew up and started to breed amongst themselves, they passed on these gentler traits. It might then have been only a matter of time—a few thousand years, perhaps—before early man began to selectively breed future generations of puppies for attractive or useful physical traits, such as depth of chest, powerful hind legs, and maybe even intelligence. Then, in the mass migrations of peoples over the millennia, each group brought along its own variation of the original wolf. Centuries of breeding have given us the different dogs we have today.

"How then did we arrive at the present state of affairs with bush dogs?" I challenged Roger.

"Well," he said, as he looked up to contemplate the ceiling of the restaurant, "if you took a male and female of each breed of dog known to man—and there are about eight hundred and fifty of them all told—and turned them loose in Yankee Stadium, and came back maybe twenty years later, you would probably find that all the youngest puppies were bush dogs."

I saw what Roger was driving at, a sort of reversal of the genetic inbreeding carried out by man, but the genes could never be put back together again in quite the original combination to make a wolf, because many of those genes were lost and gone forever. You would, however, be left with your basic bush dog.

Miss June's yard was like a small-scale Noah's ark. The chickens continued their scurrying around; I now saw two cats crossing the grass; and the puppies bubbled over in doggy effervescence. Not one type of animal interfered with another; all lived in perfect harmony, it seemed.

Anyone who has ever tried to photograph a bunch of mischievous puppies knows what a frustrating experience it can be. Molly's little puppies were all over the place and wouldn't oblige me by sitting still even for a moment. By the time I had set my camera to the correct f-stop, found the perfect focus, and composed the best artistic balance, my subjects had long since gone off to sniff the nearest bush or squat under it.

As I crouched and gyrated to get the best pictures, I noticed one little puppy standing in the deep drainage groove running alongside the front of the house (a result of the runoff of rains cascading from the corrugated tin roof that overhung the verandah). I hadn't seen this little one before, I was certain. I made a quick recount . . . five, six, seven, and, yes, there were eight, not seven as I had first thought. It was female, I could see that right away, but she bore hardly any resemblance to the rest of the stocky pups of the litter, who rolled over one another in nonstop play or ran to their mother to nurse, the very image of energy and vigor. This puppy looked like a shrunken, wizened replica of an adult dog on her last legs. She was half the size of her brothers and sisters, hardly bigger than a plump guinea pig, excluding her scraggly little legs, her head slightly downcast and tilted to one side, as if it were too heavy for the muscles of her neck. The very weight of the world seemed to be on her bony little shoulders. She stood with her legs spread wide apart to give herself balance and stability, but her little body tremored continuously, as though she would fall over if she made the slightest wrong movement. She had the bloated belly characteristic of a heavy infestation with parasitic worms, and her pencil-thin tail stood erect, too stiff with soil to go back down. When she would decide to take a step forward, she would actually step backwards a pace, then readjust her angle to point her body in the new direction, before lumbering forward with the cumbersome grace of a World War I military tank. The lids of both her eyes were matted closed with pus, and her left eye bulged beneath its lids, giving her head a distinctly lopsided appearance. Even so, she looked more like her mother than any of the other puppies. She was white, except for black ears and a broad black patch on either side of her head and neck, which left an irregular white line running down the middle of her forehead. She had a tiny almond-shaped brown spot above each eye, the only feature that she seemed not to have inherited from her mother. She also had a small irregular black patch on her bottom, to the left side of her tail, but this looked more like a soil stain than a haircoat marking.

"What about this little one?" I asked Miss June. "I don't remember seeing her before."

"Oh, she's the runt of the litter," she replied. "She was

born long after the rest, and she can't compete with the others for a nipple." My heart went out to this pathetic little mite of a puppy, and I bent down to scoop her up to my chest. She felt like a feather in my hand and I realized she didn't have the dank characteristic sour-milk smell of a normal newborn pup. Her tongue and gums were marble white, not the bright flesh pink of the other puppies, but I wasn't sure whether this was because of anemia or shock. Her little black nose was clogged and bubbling with cream-colored pus, and the skin of her muzzle and lips was scalded raw pink by the continuous discharge. Her bony little bottom was also raw from the constant flow of diarrhea. I held her up in my hand, pressing her chest close to my ear while I closed off my other ear, the best I could do without a stethoscope. Even above the rustle of the breeze, I could hear the sounds of her breathing, interrupted now and then by faint grumbling cries, and I could hear the bubbling and feel the gurgling of her lungs beneath her ribs. This little thing was in bad, bad shape. She definitely had bronchitis, possibly pneumonia.

Runt puppies are usually the last to be born, sometimes as much as a day after the rest of the litter. Growing from an egg that has had the misfortune to implant itself at the very top end of one or other horn of the uterus, far removed from the best blood supply, the runt is undernourished throughout the pregnancy, and is often so weak that it dies from exhaustion during the course of being born or shortly thereafter. This little one had survived three and a half weeks, but I could hardly fathom how. "She takes a little milk from the saucer," Miss June assured me, "and we have tried to get Molly to stand for her so that she has a better chance to suckle. But she soon gets pushed out of the way by the other puppies."

"Let's have a go now," I suggested to Miss June. "You get Molly to stand and I'll hold the puppy up to a nipple." It worked for a few seconds, the puppy sucking with surprising vigor, but Molly soon became agitated by our interference, and the other puppies began to crowd in to see what was going on, desperate not to miss out on anything. In the process, the little runt was quickly pushed out of the way.

With that, Miss June invited us into the house for tea. "We can give the little puppy some milk in a saucer," she added.

In every detail, Miss June's abode resembled a doll's house. Bookshelves lined two walls of the tiny sitting room. Small potted plants, framed photographs, and quotations from the Bible adorned the other walls. Every space in this minute room was well planned; every crook and cranny contained some treasure. Miss Maisy soon emerged from the minuscule kitchen beyond the living room with a large pot of tea, a plate of Scottish shortbread, and a saucer of milk for the puppy.

We began to talk; Marie-Paule to explain about her teaching, me to say I was a vet. Miss June told us that she had been born in Jamaica, sometime around the beginning of the Second World War. Both her parents were English, and had met, fallen in love, and married in Jamaica, after they had shipped out from England to work as civil servants in the British colonial government. We went on to discover that Miss June was married to Frank Pringle, a former minister of tourism, and now a financial advisor to the government of Jamaica. His parents had been wealthy and influential landowners in colonial times. Well, well, I thought, Molly comes from a very prestigious background.

"And what about Molly's story?" I asked Miss June.

"Well, let me see," she said, with a faraway look in her

eyes. "Molly is about four years old. She was adopted when she was a puppy by Joan Marshall, the Hollywood film star. Miss Marshall had left America and come to Jamaica when she discovered that she was dying of cancer. She loved Jamaica so, she wanted to spend her last days in peace. She and I became close friends, and after she died I took care of Molly. This is Molly's second litter of puppies. She had no problem at all with the first," Miss June continued with pride, "and I found homes for them all. These puppies are all spoken for, too," she added, and began to describe to what type of home each would go. "The little runt is going to a farmer who has a mentally retarded three-year-old son. He thought it would be good for them to grow up together, seeing as they both have problems. He's such a kind man, and I know he will give the puppy a good home."

"And what about the father of the puppies?" I asked.

"Oh, that's Shaba," Miss June replied with a laugh. "He's a real character. All the other dogs in the area are in awe of him. You may have seen him around the villa where you're staying. He's got his eye on the female that lives next door."

Shaba—that's a regal-sounding African name if ever I heard one, I thought. I did recall seeing him earlier in the afternoon, playing and carousing with the female in the next house. With his cocky, self-assertive manner, he was every bit the bush dog. He had a stockier body than Molly, and was a pale biscuit brown all over, except for a small swatch of white on his chest. So that's where the nutmeg spots above the little runt's eyes come from, and the brown markings on some of the other puppies, I realized.

As I sat on my little chair in the corner, with the runt puppy held across my lap, deeply engrossed in Miss June's stories, I noticed a flea making its rickety, arduous course through the hair on the puppy's neck. Trying not to let Miss June see what I was doing—for owners are often very embarrassed to be told that their animals have fleas—I tried to snap the flea between my two thumbnails. Then a second appeared, followed immediately by three more, crisscrossing one another's paths. In no time at all, I realized that the puppy was absolutely infested with fleas. I began to feel itchy all over, wondering whether some of the fleas hadn't made it into my clothing.

During my years as a vet in private practice, my axiom to dog and cat owners had always been that if you see one flea, there are nine that you haven't seen. By this calculation, I estimated that there must have been at least two hundred fleas on this little thing's body. If I spent all day at it, I couldn't hope to squash more than a few between my thumbnails. I have always been impressed by a flea's ability to learn very quickly that something, or someone, is after him. While you might be quite deft at catching the first two or three fleas, the rest seem to know you're after them, and they lie low, as if a warning were being transmitted by jungle drums. Perhaps, I thought, if we took the puppy back to the villa for a few hours, we could get some sustained nourishment into her, and at the same time I could continue the hunt for fleas. I remembered that Marie-Paule kept a fantastic pair of broad-bladed tweezers in her toilet bag that would be ideal for the purpose.

"Miss June," I said, "would you mind if I take the little puppy back with us to the villa for a couple of hours? I could really concentrate on getting some milk into her. I'll bring her back later on tonight." Miss June seemed quite happy at this and gave us a small bag of milk powder that she had been using for supplemental feeding. Bidding our goodbyes, Marie-Paule and I, with the pup wrapped in a small piece of toweling, set off back to our villa, this time by the quicker back way, to scale the coral wall once again.

Miss Polly, the young woman who looked after us in the villa, was in the kitchen when we returned. She was cooking the evening meal and tidying up, all part of the Jamaican holiday package. When she saw the puppy in my arms she immediately took pity on her and set about making a thick chicken soup, "the best nourishment," she said, "for weak children and animals alike."

While Miss Polly prepared the broth, I went straight away to get Marie-Paule's tweezers from her toilet bag and then laid the puppy out on a thick layer of several issues of the *Gleaner,* the daily Jamaican newspaper. As weak as she was, I had no difficulty getting her to stay on her back while I searched for fleas on her abdomen. I immediately saw three or four fleas zigzagging across the open prairies of her bloated tummy, but I found them even harder to catch in this exposed position than those wandering amongst the dense forest of hair on her back and neck. After half an hour of hunting and squashing, I realized the hopelessness of the task. What I needed was some good flea powder, but where would I find some at this late hour?

Try I would, though, and I left the villa to walk the narrow, winding lane toward the village. I tried to recall what sorts of shops I had seen. There were only two, if I remembered correctly, a tiny supermarket and an engine-parts shop, which never seemed to be open. I couldn't imagine either selling flea powder.

On the way to the village, I saw an elderly man leaning on his garden gate, gazing at the sunset. He was a fisherman and had returned for the evening after the day's catch. "Good evening," I said to him, "do you have any dogs?" "Yes," he replied. "Do they ever get fleas?" I inquired. "Sometimes," he answered with a quizzical expression beginning to show on his face. "Do you ever use powder to treat them?" I asked eagerly, quickly explaining the reason for my inquiries. "Oh, yes," he said with a chuckle. "Wait awhile, I think I may have some in the shed." With that, he called to his wife, who appeared at the side door of the tiny cottage, and the two of them disappeared into a garden shed at the back. I could hear them rummaging through boxes, and finally they reappeared at the door, beaming with success.

After thanking the fisherman and his wife, back to the villa I went, and with a metal comb in one hand and the canister of flea powder in the other, I set to work, shaking and combing, combing and shaking. This was great stuff, the strongest flea powder I had ever come across, probably too toxic to be sold on the American market. The label on the canister was so worn and tattered that I couldn't make out any of the ingredients listed. Within minutes, I had a thin carpet of dead and dying fleas on the newspaper. My original estimate was no exaggeration; I had killed at least two hundred of them. I continued to comb out the pup's hair, not only to remove the fleas, but to rid her of as much of the powder residue as I could. It is really quite dangerous using such toxic compounds on very young animals, all the more so if they are already ill, because the chemicals are absorbed rapidly through the skin into the bloodstream and can then attack the organs and nervous system. I was afraid that if this happened to the puppy, she could go into seizures, and maybe even suffer permanent damage to her brain or liver.

Within twenty minutes, I swear not a single flea remained on that little thing's body; I felt quite proud of myself. I placed the puppy to one side on a blanket, and with the greatest of care I folded the newspapers so as not to spill any of the contents onto the floor and disposed of the package in the garbage can that stood outside the back door of the villa. I wondered what the reaction of the owner of the villa would be, somewhere off in his home in America, if he knew of the odd practices going on in his absence.

What next, I wondered as I looked down at the pup asleep on the blanket. She needed a bath to get rid of the flea droppings that littered her coat like a dusting of fine black sand and to wash off the mats of dried diarrhea on the hair of her bottom and down the backs of her legs; she would surely feel better being clean. With a bowl of warm water and a gentle shampoo, Marie-Paule and I set about washing and rinsing, a section at a time, so as not to let the puppy get too cold, going over her hair with a blow dryer in rapid strokes. After ten minutes, the pup looked all fluffed up and shiny as a new pin. But I was still concerned about the infection in her eyes. I had no antibiotic ointments to treat her with, and every time I tried to pry her eyelids gently apart, she screamed in pain. The only remedy I could think of was to bathe her eyes in warm milk, a trick my mother had taught me from her days of treating horses in Ireland when she was a girl. The fats in the milk helped to soften and dissolve the hard-crusted pus.

In the meantime, Miss Polly had finished preparing her thick soup, and she poured some into a saucer on the stone floor. The little pup made straight for it, plunging her little muzzle into the viscid brew and lapping as though there were no tomorrow. She could hardly keep her balance on her splayed-out paws, so eager she was to drink, and by the time she had finished, her raw pink muzzle was covered in congealed, cream-colored broth, and spattered droplets stuck all over her face and whiskers.

We were encouraged by the puppy's progress, and I had killed all the fleas, but now what? If we gave her back to her mother she would end up in the nest, made of whatever, wedged in the crawl space beneath the foundation of Miss June's house, along with all the other flea-ridden pups. She would be infested again in no time at all, and we would have achieved nothing. I was also very concerned about her lungs, and the possibility that she might have early pneumonia. The very least we had to do was keep her until the next day and assure that she got sustained nourishment at regular intervals throughout the night.

Chapter Three

■ ■ ■

MARIE-PAULE HAD BECOME an old hand at giving special care to weak and sickly animals. This is all the more surprising to me when I stop to think that she came from a family background completely devoid of any experience with animals; she had not even owned a goldfish when she was a child in France, and the closest she had ever come to a dog was when one had bitten her on the bottom when she was twelve years old.

All that was to change, when we got married and lived in Scotland shortly after I qualified as a veterinarian. Between her duties as a schoolteacher, Marie-Paule would often accompany me on my calls to the poor farms of Ayshire, the countryside that Robert Burns knew so well, and where, even to this day, the rich Scots dialect is still the lingua franca. She helped me deliver many calves by cesarean, sticking by me for the long hours during bitter winter nights in drafty barns, or out on the ice-covered mountainsides in howling gales, passing me surgical instruments as I needed them. Often the struggle seemed hopeless, yet the miracle would happen, and despite my fatigue and ice-numbed fingers, I would make the final pull, and out would emerge a wet, steaming calf, a faint heartbeat barely detectable through his chest wall. Marie-Paule would help me hoist the slippery calf, weighing one hundred pounds or more, into the air by the hind legs, to spin it in circles to drain the birth fluids and mucus from its throat and nose. During lambing season, when my hands, though small and slender for a man, were still too big to feel my way inside a scrawny Blackface sheep, Marie-Paule would insert her own butter-smarmed hand gently into the ewe to describe what part of the unborn lamb she could feel, and I would instruct her how to straighten out the legs or turn the head so that she could complete the delivery for me. The reward? The thrill of seeing new life, witnessing the calf or lamb's first tottering efforts to stand—the mother, in spite of her own exhaustion, turning her head to lick and nudge her newborn. Marie-Paule and I would never tire of this miracle, I knew; no matter how many times in my life I would witness such events I would never fail to be awed by the sheer beauty of it all.

Since those days, Marie-Paule has reared many animals in need—orphaned newborn goats, abandoned dogs and kittens, a whole slew of baby chimpanzees, and even two infant rhesus monkeys. Animals seem to trust her instinctively. Maybe it's her calm, easy way of moving, with none of the jerkiness that so distresses an already frightened animal. Or maybe it's her thick French accent that reassures them, just as it does me.

The hardest time my wife has ever had with a newborn animal was rearing Gabby, a premature infant chimpanzee that I brought home from the lab one Sunday morning. Although I had already worked for ten years with various types of monkeys, I was still pretty much a newcomer to chimpanzees in those days. This was only the second baby chimp to be born since I had started working at LEMSIP six weeks earlier, the first having been born just the day before.

I discovered the birth of the baby chimp within minutes of arriving at the lab for my weekend rounds, and put out a call on the intercom for Jim, one of the technicians on duty, to come and assist me. The event was quite unexpected. As a quick check of the breeding records confirmed, the mother, Aphrodite, a sweet old female, wasn't due for another two months. She now sat forlornly in the back of her cage, the newborn straddled motionless across the palm of her upraised hand like a limp dishrag.

Jim, hardly more than eighteen years old, with an unruly mop of yellow hair hanging over his eyes, was all apanic on seeing the dead baby in Aphrodite's hand. "I can't believe it," he stammered. "I was talking to her not twenty minutes ago, and she gave no signs of being in labor."

"Don't get upset," I said, trying to calm him down. "It's quite often that way with animals: the babies just pop out unexpectedly." Jim pulled himself together and set about preparing some syringes and needles to anesthetize Aphrodite so that we could take the lifeless infant from her. Eight weeks premature for a chimpanzee is equivalent to ten weeks or more for a human baby. This little thing was a flimsy, almost hairless wisp, and once we got her away from her mother, I could detect no signs of life in her. Laying the infant's ice-cold body aside on a nearby table, I turned my attentions to the mother, who lay anesthetized in the cage. I wanted to make sure Aphrodite had passed all of the placenta, and that no pieces remained inside her to set up an infection in her uterus. I also needed to take a blood sample from her to make sure she didn't have a preexisting infection that might account for the miscarriage.

When, some minutes later, I turned back to the stillborn infant on the table, I couldn't believe my eyes. Beneath the thin shiny skin of her chest, I saw the faintest fluttering of a tiny heart, although she still hadn't started to breathe. Her gums and tongue were a dirty, deep purple, a clear indication of oxygen deprivation. Then began the desperate effort to get Gabby breathing and raise her body temperature. While Jim stretched her tiny neck out and pried her mouth open, I managed, with some difficulty, to insert a kitten-size breathing tube into her trachea. Jim started to deliver a gentle flow of oxygen from a small cylinder while I rhythmically compressed and released her chest wall. Jim then ran to get a bucket of warm water, and we immersed the little body up to the neck to raise her body temperature, while continuing with our resuscitation efforts. To my amazement, after only five minutes or so of struggle, the infant took one deep breath on her own, and soon she was going full speed ahead, with no further assistance from me. Gabby's gums and tongue immediately changed from the dirty, deep purple they had been and took on the bright pink color indicative of good oxygen supply. Jim was thrilled, and very proud of himself, as well he should have been: he had performed like a seasoned pro. Even with this dramatic change, however, it would require a miracle for this little thing to survive.

In those days, LEMSIP didn't yet have its well-organized nursery, staffed by skilled and dedicated caregivers. (In fact, it would take me several years to develop one and to find the right people to staff it.) Consequently, I had no alternative but to take the little baby home for rearing, along with an incubator, several small oxygen cylinders, and an array of emergency drugs, intravenous catheters, feeding tubes, and other medical supplies.

This was where Marie-Paule came in. In the beginning, I was very good about alternating Gabby's hourly feedings with her, round the clock, but after two nights in a row, I was so exhausted I didn't even hear the baby cry in the night or Marie-Paule wake up to prepare the milk for her. Marie-Paule carried the load entirely on her own, I must shamefully confess, for the next six days and nights.

My bringing home baby chimps to be reared was not always fair on the family, especially for Marie-Paule. After all, she put a lot of effort into their rearing, and developed a deep sense of attachment for them all. Gabby was special, however. She was like the most delicate little flower. It was touch and go whether she would make it, in spite of the intensive care. Marie-Paule even began to question my motives: how could I, she wondered, want to help these animals, only to return them to the lab, to a lifetime of research. When Gabby reached ten months of age, she developed meningitis and died from its aftereffects. It was a heartbreak that Marie-Paule would never quite get over.

Chapter Four

■ ■ ■

At least, I thought, this little Jamaican puppy wouldn't be as difficult to look after as Gabby. Miss Polly had magically produced a fine-woven wicker basket, and with a good supply of soft towels to make a warm nest for the puppy, Marie-Paule and I went to bed early, the alarm clock set to awaken us every hour for feedings. But each time we got out of bed, we found the puppy had somehow crawled out of the warm basket and had ended up asleep on the ice-cold stone floor, snuggled alongside my tennis shoes, which, no matter where I had left them, she seemed to find, homing in on them like a beacon. Puddles of urine and splatters of diarrhea close by told the story. What a courageous little thing this puppy was. Barely three and a half weeks old, unable to see, bled almost to death by fleas, breathing raggedly with bronchitis, parasites ravaging her intestines, she still could not bring herself to soil her bed.

Although this was the tropics, and the nights seemed sweltering to us, a small creature like this puppy couldn't hope to maintain her body temperature on her own. She didn't have the warmth her littermates and her mother provided, and, in all likelihood, the heat-regulating mechanism in her brain wasn't fully functional at this early stage in life. Besides, the smaller the creature, the greater the heat loss through the skin because the ratio of the surface area of skin to body weight is so large. A tiny mouse, for example, loses body heat much faster than a bloody great elephant, because the mouse has a proportionally larger skin surface than the giant pachyderm. This would never do, we realized. There was nothing for it but to take the puppy directly to bed with us, to keep her warm.

I donned a pullover, despite the warmth of the night, and snuck the little puppy down between the wool and my bare chest. With her every movement, she dug her sharp little claws into the skin of my chest and swiped her wet nose back and forth across my throat. When she did finally fall into deep sleep and stopped moving and making whimpering and grumbling sounds, I found myself continually checking to reassure myself that her little heart was still beating. There was no possibility of sound sleep; neither Marie-Paule nor I would get any real rest at this rate. Halfway through the night, we ended up taking separate bedrooms, hoping to afford each other at least some respite between the rotation of hourly feedings and clean up. By the first light of dawn, we were exhausted. My eyelids felt as if someone had poured wet cement into them, and Marie-Paule's eyes were puffed and as red as beets. The night had been an exercise in "two steps forward, one step back," as the puppy consumed alternate feedings of milk and chicken soup, only to vomit between feedings or have another bout of diarrhea. On top of everything, her bronchitis seemed only marginally better. This was turning into some vacation.

As the next day progressed it became obvious that we couldn't give the puppy back to Miss June in the state she was in. The puppy needed a great deal more careful nursing if she was going to pull through.

Marie-Paule and I continued the hourly feedings of milk and chicken broth, and the puppy even began to take small mouthfuls of finely minced beef liver and chicken. We alternated baby-sitting duties throughout the day, to give each other a chance to relax. One of us always stayed behind to take care of the puppy and make sure she didn't wander out of the house through the sliding glass doors. I had visions of her falling into the swimming pool, or even tumbling down the side of the cliff. When it was my turn on duty, it seemed I spent much of my time following her around with a wad of paper towels. Even though the floors of the villa were made of stone flags, I knew the urine and diarrhea would seep into the cement seams in between to make a lasting stain.

In spite of the heat of the mid-morning sun, I was surprised to find how difficult it was to keep her warm. If I left her for more than a few minutes alone on the floor, her body would start to shiver. Wrapping her in towels so that she could sleep in Miss Polly's basket wasn't very successful either, because she would squirm her way out when my back was turned. I would find that she had wandered off, to sit miserably in some corner of the living room or along the hallway leading to the bedrooms. Carrying her around inside my shirt, with her little head poking out, was hardly more successful, because she constantly struggled for me to put her down so she could take a walk. Puppy-sitting was becoming a frustrating and full-time occupation. I was thankful for Marie-Paule's return at midday, and after a little spot of lunch she said it was her turn to take care of "Pup," and that I should take some time off for myself.

For me, one of the greatest pleasures of Jamaican vacations is exploring the countryside on my own, following each little lane to trace where it might lead, discovering the mountain villages, with their collections of small, brightly painted houses and neatly tended gardens, and seeing the luxuriant displays of tropical flowers all along my route.

Even after twenty-four years of living in the United States, one of the things I miss most (besides the atmosphere of a village pub and a wholesome jar of bitter or black porter) is the feeling of walking along an English country lane, or boreen, as they say in Ireland. I doubt that there is anything quite like it anywhere else in the world. You never know what to expect as you round the next bend, or pass the next hedge. Everywhere you hear the background twitter of blue tits as they nervously flit amongst the ancient gnarled limbs of hawthorn and yew, hazel and wild woodbine.

A Jamaican lane is quite different, of course—more vibrant than an English lane, and with its own very special charm. Around each turn, the narrow road reveals seemingly endless vistas of tiny terraced fields of bananas, corn, and sweet potatoes, mustard and aloe, dotted here and there with stands of coconut palms and, at higher elevations, the coffee plantations. Every square foot of rich red earth seems to be used to maximum effect.

On this particular afternoon, I drove up into the mountains, following a winding road that got narrower and narrower as I went, the surface changing from potholes, the size of which would put New York City's finest to shame, to downright disintegration, ending finally as a grass track. I had passed a breadfruit plantation, where I had noticed a group of shirtless men wielding machetes as they trimmed the trees and harvested the fruits. I was determined to drive as far as I could before being forced to continue on foot. The terrain ahead was beginning to turn into lush tropical rain forest, and I was sure I could get in some good bird-watching, perhaps see some Jamaican lizard cuckoos up close, or tiny todies, with their lustrous green backs and brilliant ruby red throats. Abruptly, the car rolled to a stop in front of a patch of exposed limestone. I studied the crevices in the smooth rock surface before me. Having calculated precisely how to steer the car to miss the holes, I cautiously released the clutch, gently pressed down on the accelerator, and—*whoomph*!—slipped into a crack not two feet in front of me. The nose of the car was pointing sharply down, the trunk suspended in the air. Holding my breath, I very gently opened the driver's door and slid out. I couldn't believe my eyes: both rear wheels of the car were in the air, and the front end of the chassis was lodged in the crevice against the rock. Holy Mother of God! How could I be so stupid? I silently lamented. The nearest telephone had to be at least thirty miles away, and I felt pretty sure the nearest tow truck would be no closer. I considered whether I might have to abandon the car and hike back down the mountain to the nearest community in the hope that I would find someone kind enough to give me a lift to a service station. Then I remembered the men I had seen working in the breadfruit plantation. There was no choice but to walk back down the mountain and try to solicit their help, although I wasn't sure what reception I might get from them.

I need not have been concerned; a dozen of them immediately dropped their tools and followed me back up the mountainside to the car. The way the men gathered around the car, I realized that they intended to lift it, not push it, as

I had expected. "Don't do that," I said to them, "it's going to be too heavy, and you'll hurt your backs." "No problem, mahn," one of them said to me with a grin. "You just sit in da car and be ready to turn da steerin' wheel if we tell you." With that, half a dozen of the men grabbed the car under the front bumper and began to lift, while the others took hold of the back, ready to lift once the front end had cleared the crevice. After much straining, they let out a roaring cheer as they set the car back down squarely on her four wheels. I tried to give them some money, but one of the young men leaned in through the driver's window and indignantly said, "We didn' do dis for da money, mahn!" "Well, I think it deserves a Red Stripe beer all around," I announced. Very happy with themselves, they clambered onto the hood, held on to the posts of the open doors, or squeezed on top of one another in the car seats. With the greatest of caution, I managed to turn the car around, and I drove ever so slowly back down the mountain to the first village bar to celebrate.

When I arrived back at the villa in the late afternoon I found Marie-Paule playing with the puppy. She filled me in on all the little progresses the puppy had made while I was away, and how she had given the first signs of happy tail-wagging. But I couldn't help thinking that in a few days we would have to leave Jamaica, and without some really big breakthrough, we would be abandoning this little puppy at the very time she most needed us. There was a limit to how quickly I could expect her health to improve. Admittedly, I had ridded her of her fleas, but as soon as we gave her back to her mother, she would be reinfested in no time at all. Her diarrhea could turn into a life-threatening condition at any moment, and the infection in her eyes was of grave concern to me. Even with constant bathing of her eyes, we had managed to get only her right eye open; the lids of her left eye remained solidly shut. Although she seemed not to have bronchitis or pneumonia, as I had first thought—the infection turned out to be confined to her nose and sinuses—I was still very concerned about her anemia, caused mainly by fleas. It seems hard to imagine that a creature as minute as a flea can jeopardize the life of its much larger host. Yet multiply the effects of the two hundred fleas I estimated the puppy had, probably since the moment she was born, and bear in mind that as soon as each flea was sated it would hop off to lay its eggs in the bedding, its place immediately taken by another flea who had already laid its eggs, and a picture of cataclysmic bloodletting emerges. This little puppy, with all her other health problems, couldn't possibly generate new red blood cells fast enough to keep pace with this degree of blood loss. Even without any further flea infestation, it would take the pup's bone marrow weeks to replenish her blood.

The following day was much of the same. By Wednesday morning, after three consecutive nights of little sleep, Marie-Paule and I were at the point of no return, zombies going through the motions of pretending to be having a great time. The puppy, on the other hand, had made some real progress. She was eating regularly, and although she still had bouts of diarrhea, her vomiting had subsided. There was no mistaking that she was stronger, but she still didn't show the vitality of a normal puppy. She continued to walk with a clumsy waddle, and had to take two steps backward before she could alter course to proceed forward again. Her left eye remained shut, and she still had a significant discharge from the now-open right eye. But of greatest concern to me was the anemic pallidness of her gums and tongue. What the puppy really needed, I began to realize, was a blood transfusion. No amount of feeding could make up for her degree of anemia. But this was Jamaica. What chance did I have of finding a vet who could—or would—perform a transfusion on a scrawny three-and-a-half-week-old pup?

I dug out the telephone directory for the Island of Jamaica from the pile of books and magazines that lay strewn across the coffee table and turned to the section for veterinarians. To my consternation, there was only one vet listed for whole of the island, and he was located in Montego Bay, an arduous two-hour drive north of Treasure Beach. I knew there was certainly more than one veterinarian in Jamaica, but without names to go by, there was no way I could find their addresses and telephone numbers. I considered asking our neighbors if they might know of a vet, but this was mainly an agricultural area, and I didn't think my chances were great of finding a local rural practitioner who would be willing or equipped to perform a transfusion on a puppy. What to do? I wondered.

I sat bleary-eyed, looking down at the pup asleep on my lap, as I drowsily drank my early morning coffee, and considered my options. If I couldn't find a veterinarian to perform the transfusion on the puppy, could I entertain doing it myself? That would be ridiculous, I thought. Marie-Paule and I were on vacation, just a short eight days of relaxation, including the two days of journey to and from Jamaica. It was already Wednesday, and we would have to leave Calabash Bay quite early in the morning on the coming Saturday in order to catch the flight back to the States. How would I even begin to go about getting the materials I would need to perform a transfusion? What about the risks involved in performing the procedure on such a tiny, sickly puppy?

I picked her up in one hand to hoist her gently up and

down in the air, trying to get an impression of how much she weighed, as though I were estimating the weight of a bag of flour. After nearly thirty years of working in research, I had become accustomed to estimating weights of monkeys and chimps in kilograms, and blood volumes in milliliters, but when it comes to dogs and cats, I can think only in terms of pounds and fluid ounces, a throwback, no doubt, to my days as a country vet in Scotland. I would have to make some careful calculations of body weight in pounds and then convert to kilograms to work out the ratio of blood volume in milliliters.

Let me see now, I said to myself. A baby chimp weighs on average one and a half kilograms at birth; this little thing certainly doesn't weigh that much. I continued with my mental calculation as I hoisted her up and down again for good measure. I'm sure she weighs less than two pounds, I pondered, which is equivalent to a good bit less than one kilogram. I made some quick pencil calculations on the edge of a nearby sheet of newspaper. I tried to recall what Dr. Wojtek Socha, my colleague at LEMSIP and head of the blood group immunology and genetics laboratory, had often instructed me: The total blood volume of a human being or an animal is approximately seven percent of the body weight. So if I assume for simplicity's sake this little puppy's body weight is one thousand grams, then her blood volume must be about seventy milliliters or so. I double-checked my figures to make sure I had the decimal point in the right place. Could I now make a reasonable guess as to how anemic she is? I had certainly never seen an animal with such pale gums and tongue, and I had seen some pretty bad cases in my time. She had to have lost well over half her red blood cells; more like three-quarters. Let's assume, I continued, that a normal puppy's packed cell volume is 40 percent. The packed cell volume, or hematocrit as it is also called, is the fraction of liquid blood taken up by the red blood cells, the remainder being the plasma. That would mean that I would have to replace somewhere between thirty-five and fifty milliliters of blood. But wait a minute; I remembered Dr. Socha cautioning me on many occasions never to replace all the blood that an animal has lost in a single quick transfusion, because this runs the risk of suddenly overloading the animal's vascular system. To give this little puppy all the blood she needs in one go would take me at least two hours, and I could never hope to keep her still for that length of time without her struggling and dislodging the intravenous catheter. If I gave the blood any faster, I would run the risk of sending the puppy into heart failure. This would lead to her developing pulmonary edema, and she would ultimately drown in her own body fluids. Handling a puppy as tiny as this, and with such severe medical problems, would certainly not be as easy for me as for a doctor attending a human patient who lies cooperatively in a hospital bed where the blood can be given slowly, drop by drop, over twelve or even twenty-four hours. No, I said to myself, it would be safer if I underdosed the puppy with blood, rather than risk giving her too much.

And who would I use as the blood donor? Dogs have over ninety blood groups, that much I knew, but most importantly, dogs fall into two major categories: those that are positive, and those that are negative for what is called hemolytic blood factor A. Only A-negative dogs, like their human type O counterparts, are suitable as universal blood donors. If you transfuse blood from an A-positive dog into an A-negative dog, an immediate immune reaction called anaphylactic shock occurs, and the dog receiving the blood dies from massive destruction of its own, already depleted red blood cells. But blood group immunogenetics was not exactly my field of expertise. Say I transfuse the puppy with blood of the wrong type, I continued to argue with myself; she would then die from the mismatch. Should I choose her mother's blood or her father's? Presumably Molly's blood would be compatible with her own puppy's, otherwise the puppy would have already died from the dog equivalent of human Rh-factor incompatibility. Most, if not all, of the antibodies that provide a puppy with immunity against the most common diseases are passed in the colostrum, the first sludgy milk that the mother secretes, and not through the placenta, as in human beings and nonhuman primates. If the puppy had inherited a blood factor from her father that was incompatible with the mother's system, the puppy would be affected only after she was born, once she started to nurse from her mother. Could just such a mild incompatibility have played a role in the puppy's severe retardation in physical development? If this was so, I should perhaps consider using the father, Shaba, as the blood donor. I had to be practical, though; not only would it be unlikely that I would ever be able to find him, but he would almost certainly not take kindly to my attempting to take blood from him.

No, I said to myself, you're getting carried away in silly theory; it would be better to stick with Molly, the pup's mother.

IF ONLY I COULD give a quick call to Dr. Socha in his lab at LEMSIP and ask his advice. One of those rare people you feel so much the better for having known, Dr. Wojtek Socha is the quintessential white-haired gentleman that we all tend to imagine when we think of a university professor. He studied under the famous Alexander Wiener when he first came to America from Poland in the early 1960s. Wiener in turn was the student of the even more famous Karl Landsteiner, who discovered the existence of blood groups in the 1920s. Wiener and Landsteiner codiscovered the Rh factor in human beings, which made possible the twentieth century's single greatest step in improving pediatric care, saving the lives of hundreds of thousands of newborn children who otherwise would have died from jaundice, anemia, and brain damage.

For all his brilliance, Dr. Socha is an unpretentious and rather self-effacing man, always willing to share his expertise, and has given me invaluable help on a number of occasions over the years when I have had problems in the field. But then I remembered he would not be there; he was off lecturing at the College de France, in Paris, for the summer. I would have to think this through on my own, I realized. But even with my most careful planning, I would have to accept that transfusing the puppy would not be without risk. The more I thought about it, though, the more I realized I had no alternative but to try.

Chapter Five

■ ■ ■

Attempting to save an animal in distress requires a considerable amount of trying—nothing comes easily. Success requires patience and commitment, without which even the most well intentioned can fail.

Rebel was a blue-black Labrador retriever puppy, the first animal Marie-Paule and I ever adopted, not long after we were married in those early days in Scotland.

One Sunday morning, around eight o'clock, I heard a knock on the door of the clinic where I worked. A primly dressed, rather austere-looking middle-aged woman stood at the door, an effervescent six-week-old puppy at her side, straining on his lead. "May I help you?" I asked the lady. With the rolling R's that only a Scottish tongue can make its way around, she announced, "I'm sorry to disturb you so early in the morning, but I'm wondering if you would put my dog to sleep."

"Why do you want him put to sleep?" I asked.

"He has a bad ear infection," she replied, "and try as I may, I don't seem to able to get on top of it."

I lifted the puppy up onto the examination table. Sure enough, he did have an infection in both ears, caused by mites. The ear canals were inflamed, and there was the heavy, acrid smell that comes with infection. "Ah, but that's no real problem," I said. "I can clean his ears easily enough, and I'll give you some antibiotic ointment to apply twice a day, and in a few days he'll be as right as rain."

To my surprise, the lady seemed not the least bit pleased by my positive prognosis. "Whhell!" she replied, in her clippped Scots accent, a troubled expression beginning to loom on her face. "Whhell!" she began again, "I'm a schoolteacher, you see; I work in the primary school in Girvan, and I'm away most of the day. When I return home in the afternoon, I find the puppy has gnawed all the carpets and the cushions, and he's not the least bit house-trained, you know!"

"But all puppies are like that," I said in jocular fashion. "He'll grow out of it soon enough."

I could see by the expression on her face that, without having intended to, I was backing this straitlaced schoolmarm into a moral corner. She had obviously made up her mind, long before walking into the clinic, about what she wanted done with Rebel. Yet here she was, an educator of young children, on the Sabbath, no less, trying to talk to me into a dastardly act, just because she couldn't be honest with me. I was angry, but not surprised: some of the most callous attitudes I have experienced over the years towards animals have come, not from farmers, or even research scientists, but from pet owners. I knew I had to back off, however, or she might become so frustrated that she would storm out and take the puppy to another veterinary surgeon who might have no second thoughts about complying with her wishes.

"I understand," I said, with rising anger that I could barely control. "I'll put him to sleep *after* I've had my breakfast."

"Ooh," she replied, "but I was hoping I would be able to take the body back with me."

"Why on earth would you want his body?" I asked in amazement.

"Whhell," she continued, "how do I know that if I leave the dog with you, you won't sell him to research?"

I was dumbfounded. In as sympathetic a tone as I could muster, I assured her that I would never do such a thing. Her fears allayed, she left.

Rebel turned out to be a great dog, one of the most intelligent I have ever come across. He accompanied me as I drove along the winding mountainous roads to my farm calls, standing upright on the backseat of the car, forepaws straddled over the shoulder of the driver's seat while he chewed my ears with his needle-sharp teeth. The first time I took him to a farm, he jumped out of the car and chased all the chickens, creating total havoc and landing me in an awkward position with the farmer. I scolded Rebel royally; he never once chased a chicken again, and he could walk quietly through a flock without causing pandemonium among the birds.

In spite of what his first owner had told me, Rebel was well on the road to being housebroken. Like the little Jamaican puppy who did all she could to valiantly struggle out of her sleeping basket when she got the urge to urinate, so as not to soil her bed, Rebel would whine urgently to be let out, but he could never quite make it outdoors in time. Rebel's problem was that we lived in a flat in the keep at the very top of a small fifteenth-century castle. Penkill Castle, as it was called, was nestled in a deep glen, way out in the middle of nowhere. We had planned to get married in April, when the daffodils came into bloom, but, quite by chance, I found the apartment in December, just before Christmas. Miss Fraser-Darling, an elderly spinster and last in the family line of owners of the castle, would rent the flat out only to a married couple. Consequently, Marie-Paule and I decided to bring the wedding forward to January and move in as soon as possible.

The apartment, which had a large bedroom/sitting room (with a four-poster bed and a heavily carved oak fireplace, depicting the four saints of Scotland), a small kitchen, and an even smaller bathroom, could only be reached by a spiral staircase made of stone. The wall alongside the spiral stairs featured a hand-painted mural depicting the love life of King James I of Scotland. Each step was worn into a deep bow from the tread of countless feet over the centuries.

Poor Rebel, the urge suddenly upon him to urinate, would whimper to be let out. Then he'd career down the staircase, as fast as his legs would carry him, with me in hot pursuit in hopes of getting ahead of him to open the castle's huge oaken front door. But he could never make it out before he would let out a great deluge on the last few steps. These granite steps hadn't been peed on by a dog in their four-hundred-and-fifty-odd years of existence, and Miss Fraser-Darling was not about to let this habit go unchecked. She complained bitterly, and I was forced to part with Rebel. By good fortune, I found a kindly farmer and his wife who eagerly agreed to take him. The outdoor life of a farm dog suited Rebel—and his bladder—perfectly.

Chapter Six

■ ■ ■

I COULD NOT ALWAYS COUNT on sick animals to present themselves to me in the comfort of a well-appointed veterinarian's office or a lab clinic. I found I would sometimes have to go to them, on their terms.

One such occasion involved Lucy, the famous sign language chimpanzee. Lucy was born in the United States and spent the first eleven years of her life as a member of a human family. In his book *Lucy: Growing up Human,* Maurice K. Temerlin, a psychotherapist, describes how he, his wife, Jane, and their young son raised Lucy as though she were a human child. Not only did Lucy sleep in her own bed (with sheets), but she shared totally in the daily lives of the family. Most unusual of all, she was taught to communicate in American Sign Language. Concerned for her future, and wanting to let her live out the rest of her hopefully long life as a true chimpanzee, not stranded somewhere between being a human being and a humanized chimp, the Temerlins took

Lucy, along with a ten-year-old chimpanzee named Marion, to the Gambia. Maurice and Jane also took along with them a young American student named Janis Carter, who was studying for her master's degree in psychology at the University of Oklahoma. Their plan was to first integrate Lucy and Marion into a large group of young, wild-born chimps who had been confiscated over the years from pet owners or from smugglers and then to see if they could be introduced into the wild.

The young wild-born chimpanzees had first been assembled as a group in a wildlife preserve called Abuko, near Banjul, the capital of the Gambia, and taught the ways of the wild by a young Englishwoman, Stella Brewer, and her father, Eddie Brewer, who was chief of the wildlife conservation department in the Gambia. Stella then took the group to Niokolo Koba, a national park in neighboring Senegal, where she hoped to release them to the wild. After a couple of years of painstaking trial, error, and mishaps, touchingly recounted in her book *The Chimps of Mt. Asserik,* Stella became increasingly fearful of how the resident wild chimpanzees might react to the rehabilitants, and she finally brought her chimps back to the Gambia, this time releasing them onto the group of small riverine islands known as Baboon Islands.

Introducing Lucy and Marion to the wild turned out not to be as easy, or as quickly done, as the Temerlins had first hoped, and they were forced to abandon their efforts and leave their charges in Africa when it was time for the family to return to America a couple of weeks later. Janis, however, couldn't bring herself to leave Lucy and Marion behind, so she decided to stay on in the Gambia for six months, the time she thought it would take to teach the two older chimpanzees the most important things they needed to know about coping in the world. That was nearly twenty years ago; Janis still lives in the Gambia, her whole life having been dedicated to the chimpanzees.

I got to know Janis in a most unusual way and unpropitious manner. Eugene Linden, a freelance journalist for *Time* magazine who was writing a book, *Silent Partners,* about chimpanzees and the American Sign Language program, had visited me several times in the early 1980s at LEMSIP during the course of his research. Eugene had worked in the sign language program in Oklahoma, knew all the chimpanzees (including Nim, Ali, and Booee, who have also had books written about them), and had visited Janis Carter in the Gambia. During the course of our many conversations, I may well have expressed a dream of one day going to Africa and seeing these animals for myself, although, I must confess, I don't remember making such statements.

The next thing I knew, I received the nastiest letter you could imagine from Janis, berating me for my audacity in thinking that she would ever invite me to the Gambia to see the chimps on the Baboon Islands, when I was nothing but a user of helpless creatures, a ghoul bent on vivisection, the very essence of everything she detested. After two typewritten pages of this invective, her last sentence read "However, if I have misjudged you, perhaps we can talk about it."

We did talk about it, during a brief visit she made to the lab. Deciding that I was perhaps not the snake she had taken me for, Janis asked me to come to Africa a few months later to help her integrate two groups of chimpanzees—Stella Brewer's, and her own, which had grown considerably in number since Lucy and Marion's arrival a few years earlier. The plan was to take some of the chimps from where they were living on the smaller of the Baboon Islands and move them together onto the largest, an island twelve hundred acres in size. Janis wanted my help in making sure that neither of the two groups of animals carried any infectious disease, especially tuberculosis, which could run rampant through them.

The long, narrow Baboon Islands, located about 175 miles upstream from the ocean, extend over four miles or so of the river. Because of this, I had to set up my medical station each morning at different points along the riverbank, depending on which of the islands Janis wanted to work on that day. This in itself was quite a task. I had to follow religiously my system of laying out the equipment and supplies so that in an emergency, such as if one of the chimps went into respiratory or cardiac arrest, I could quickly lay my hands on what I needed.

My days in the Gambia were mostly spent alone. Janis wouldn't allow me to go on the islands with the rest of the team for fear that some of the older chimpanzees, recognizing me as a stranger, might attack me. I would sometimes have to wait hours on my own, with nothing to do but birdwatch or gaze at the hippos bathing on the far side of the river, while Janis and the others were off trying to dart the chimps. One such day of boring nothingness was punctuated by the sudden sound of snapping twigs and the rustle of bushes. I looked up to find that I was surrounded by a troop of baboons, sixty to eighty individuals all told. I was both thrilled and frightened to find myself the center of their curiosity, the roles reversed from what I was used to in the confines of the research laboratory. Several adult males, weighing seventy pounds or more, eyed me suspiciously through the branches. Their threatening yawns exposed two-inch-long razor-sharp canine teeth. A number of mothers, with infants at breast, grunted repeatedly like pigs to indicate their concern at my presence. Every now and again they would let out high-pitched screeches, as their infants, no longer scared by my presence, approached me ever closer in repeated forays, showing off cockily to one another, only to hightail it back again to the safety of their mothers' embrace when they found they had overstepped their bounds. If one of these babies should get too close to me and scream, I thought to myself, I'm dead meat; I'll find myself attacked by the whole troop. After several minutes of tense standoff, the troop, probably in response to some signal from the leader male that I was unable to recognize, picked up and moved off to their feeding grounds, which, I eventually realized, they did each day. The confrontation was over.

While I waited on the riverbank each day, Janis and her Senegalese assistants, René and Bruno, together with three European student volunteers, set about tranquilizing the chimpanzees on the different islands and bringing them to me for examination a few at a time. Tranquilizing the chimpanzees was often a difficult and painstaking procedure. With a little luck, and a deft flick of the wrist, Janis and her helpers could get to close quarters with the younger animals and lash out with a syringe and needle that each had hidden up a sleeve or in a trouser pocket. Sometimes their sleight of hand was so effective that a young chimp wouldn't even realize it had been injected. But with the larger chimps this approach was impossible because of the danger involved, and everyone was forced to use four-foot-long blow pipes loaded with anesthetic darts. It requires a great deal of huff and a considerable amount of skill to blow the feather-tailed darts twenty feet through the air with enough force for the dart needle to penetrate the tough skin of an adult chimp. If the adults and larger juveniles were in a group, the procedure was rendered all the more difficult because all the members of the group would have to be darted at the same moment. There would be little chance for a second go if Janis or her assistants missed the first time. Once the chimps knew what Janis and her team were up to, they would take off into the forest, and it would be nearly impossible for them to approach the animals again.

One group to be worked on—four adults on the smallest island—resisted all attempts to be darted, they had become so wary. I came up with what I thought was the brilliant idea of giving the anesthetic in bananas. I injected half a milliliter of anesthetic solution, a drop here and there, through the skin of each banana and then marked them with large black X's. In this way, the chimps wouldn't be able to smell the anesthetic, and the bitter taste would be masked by the fruit. We then cleverly mixed the treated bananas with untreated, unmarked fruits, to further fool the chimps. Despite the seeming perfection of our plan, the animals took not a single marked banana, gorging themselves only on the untampered fruits. After two tedious days of watching and waiting, we gave up; the marked bananas lay scattered and shriveled from the heat. We were forced to abandon any further efforts to work on these chimps.

Occasionally, the tension and frustration was broken by some amusing occurrence with the chimps. One afternoon a full-grown male managed to climb into Bruno's boat, which Bruno had left tied to a tree along the riverbank. The male tried several times to start the motor by pulling the cord, as no doubt he had observed Janis and the others doing on many an occasion. Fortunately, Bruno had removed the starter key. The chimp then untied the line and reclined in the stern of the little boat, his arms spread over the gunwales, while the strong current of the river set him adrift, heading downstream at great speed. All the chimp needed was a boater hat, a pipe, and a cool glass of champagne, and he would have cut the perfect picture of a Cambridge University don punting down the River Cam on a Sunday summer's afternoon.

But except for the odd moment of comic relief, darting and anesthetizing the chimps was grueling work. One of the most frightening parts of the job was dealing with four to six anesthetized chimps at the same time. One would never consider allowing this to occur in the laboratory setting, but this was the wilds of Africa, and there was simply no way of avoiding mass anesthesia, because the animals couldn't be sedated one at a time. Everyone had to be extremely observant during these times, making sure that no one animal was forgotten, even for the briefest moment, in case a problem arose, such as one getting its neck kinked, or suffocating on a great glob of saliva caught in the back of its throat, or vomiting and aspirating stomach contents into its lungs.

Once their medical examinations were completed and I had taken all the blood samples I needed, tested the chimps for tuberculosis, and examined their stool samples under the microscope to determine what types of intestinal parasites they carried, the chimps were taken to the large island, where they were allowed to recover from anesthesia in large cages constructed especially for that purpose. They were released to freedom some hours later.

The Gambia River, even this far upstream, is incredibly tidal, the water level rising and falling every twenty-four hours by four to six feet. The steep, muddy banks, most exposed at low tide, were treacherously slippery, and any false step when carrying the large, heavily sedated chimpanzees into and out of the tiny, constantly bobbing boats could result in a chimp's falling into the water and drowning. Even when conscious, chimpanzees, unlike monkeys, are unable to swim—probably because of their body configuration—and they sink like stones as soon as they hit water, without even attempting to save themselves. When it came time to capture Lucy, Marion, and the other chimps on their small island, the lunar phase had shifted so that low tide was now during daylight hours. Janis calculated that the safest time to transfer the animals from the riverbank into the boats would be midnight, when the water level would be at its highest.

Fortunately, it was a cloudless night, which would make our work easier. I could readily make out the small boat in the silvery light of a full moon, putt-putting its way upriver toward me as I waited in the dark shadows of the night, nearly devoured by hordes of mosquitoes. Even under the minimal illumination of the moon and flashlights, I could tell that something was dreadfully wrong with Lucy. She was an unusually long, thin animal, but, even taking this into account, I could immediately see that she was emaciated. Her eyes were sunken and her gums were pale, and I suspected she had severe anemia. She was having great difficulty breathing under the effects of the anesthetic agent that Janis had used to dart her, made all the worse, I was sure, by the diluteness of her blood. Janis was beside herself with worry.

I examined Lucy and took blood from her. It was thin and watery; there was no way we could return her to the islands after she awakened from anesthesia. Janis had a small transport cage close by, and we put Lucy in it to await her slow recovery from anesthesia during the remainder of the night. I was very concerned, but I would have to wait until daylight to examine her blood and determine just how anemic she was. I knew Janis would never forgive me if something terrible happened to Lucy.

Janis was so distraught she decided to spend the night

hunched on the wet ground by the open door of Lucy's cage, her long dark hair hanging over the blanket she had wrapped around her shoulders to ward off the mosquitoes. While Janis watched over Lucy, I made my way down to a small wooden dock farther along the riverbank, hung up my stethoscope on one of the supports, took off my clothes, and waded deep into the warm, fast-flowing water for a stand-up bath. It was heavenly to wash away the clammy sweat and feel my tensed muscles relaxing. Suddenly, Janis started screaming; something terrible was happening to Lucy. Clothed only in the stethoscope around my neck, which I had grabbed from the post as I hurried by, I scrambled up the slimy embankment, my toes repeatedly slipping in the mud, and rushed to Lucy's cage. I leaned in through the open door to listen to her heart and lungs. Lucy was groaning with every breath, which made it difficult for me to hear properly. "What's wrong with her?" Janis shouted at me in panic. I continued to listen, concentrating on every gurgling sound in Lucy's chest. "What's wrong with her?" Janis screamed at me again. "For God's sake, why don't you answer me?"

"Why don't you bloody well shut your bloody mouth for a moment and give me a chance to figure out what's wrong!" I roared back at Janis. She fell silent. Thirty seconds later, I breathed a sigh of relief. Lucy's heart and lungs sounded normal; she was only having a hallucinogenic reaction to the anesthetic, which would pass as the drug left her bloodstream. She should make it through the night.

With the coming of dawn, I set about completing my examination. Bert, one of the two Dutch students who worked with Janis, had rented an old bicycle for the equivalent of a dollar a day from a man he had found in one of the villages. I used this as a makeshift centrifuge, the best I could think of to take the place of the machine used in hospital laboratories to spin down blood samples to determine the hematocrit, the percentage of a blood sample taken up by the red corpuscles and a measure of the degree of a patient's anemia. I drew Lucy's blood up into the delicate glass hematocrit tubes, which measure about one-sixteenth of an inch in diameter and four inches in length, and taped them to the spokes of the bicycle wheel. Bert sat patiently on the ground, with the bicycle resting upside down on its handlebars, spinning the pedals at a constant speed. I would stop him occasionally to measure the height of the column of red blood cells that had settled at the base of each tube and compare the readings with the last I had obtained, so that I could determine when we had a final accurate measurement. The standard electrical medical centrifuge spins blood samples at about three thousand revolutions *per minute,* and it takes only ten minutes of spinning at this rate to get a final result. There was no way, of course, that poor Bert could spin the bicycle wheels at that speed, and it took him the best part of two hours before I was satisfied that we had an accurate reading. His assignment was made somewhat more bearable by a few bottles of Guinness stout, which he had somehow found in the village where he had rented the bicycle.

The result of the test was alarming: Lucy's hematocrit was only 22 percent, instead of the 40 to 45 percent of a normal chimpanzee or human being. I examined Lucy's stool sample under the microscope and found that she had an overwhelming infestation of hookworms, microscopic parasites that attach to the intestinal wall and suck the blood of the host. This was Africa, and not the convenient location of a medical laboratory like I was used to working in in America. Had Lucy lived at LEMSIP, I could have simply treated her with antiparasitic drugs and daily vitamins, and over the course of four to six weeks she would have bounced back to health without any trouble at all. But we couldn't release Lucy back onto the islands in her present state, nor could we keep her locked up in this tiny cage for the long period of recovery that she would need. And I couldn't remain in Africa long enough to fully monitor her progress. No, this situation called for unusual steps: Lucy needed a blood transfusion if she were to have any chance of surviving.

"You need to get me back to Banjul, right away," I told Janis. "Let me carry out some more tests on Lucy's blood at the Medical Research Council laboratory, and then we have to make some pretty fast decisions."

Janis drove me to Banjul in her Land Rover, a tiring five-and-a-half-hour journey over bumpy bush tracks and pothole-scarred roads. Once at the laboratory, I retested all the samples. The laboratory centrifuge gave exactly the same hematocrit reading on Lucy's blood as Bert's makeshift bicycle machine. By the time I had finished all the tests, it was three o'clock in the morning.

Janis and I set about trying to devise a plan for how we could get Lucy's blood samples to LEMSIP for Dr. Socha to type for a transfusion match. Sending the blood samples to America wouldn't be too difficult: I could send them by an express courier service come morning. The big problem was how to get the transfusion blood back from America, along with the equipment and medical supplies that I would need to perform the transfusions. There were no direct flights from the United States to the Gambia, and I might not be able to get such a large amount of material sent by courier service. I had also to take into account that this was already Wednesday, with the weekend fast approaching. This could delay the delivery of the materials, possibly by as much as a week, time we could ill afford. Then Janis remembered that she had an American friend in Banjul whose husband was on a business trip to Jamaica and would, sooner or later, be returning to Africa by way of New York and London. If we could contact the man, he might just be talked into picking up the blood for Lucy at Kennedy Airport. With any luck, we might even get the blood back by the following Sunday morning, the date of the next flight in from London.

Even though it was the very early hours of the morning, Janis phoned her friend. She readily agreed to try and help us; her husband would be returning on the Sunday flight, at five o'clock in the morning, and she would be happy to telegraph him in Jamaica.

We had done all that we could for the moment. Once the courier offices opened for business we could send Lucy's blood samples off to America for blood typing. Janis and I would then be free to return upriver to take care of Lucy and continue with the examination of the remaining chimps on the islands. But we would be out of telephone contact from that point on and could only hope that our plan would be successful.

Unbeknownst to us until long afterwards, a most complicated and remarkable series of events began. Lucy's blood arrived at LEMSIP within twenty-four hours, record time. Dr. Socha had all research stopped at the lab while the technicians busily set about drawing blood from the chimps to be typed and cross-matched with Lucy's. Janis's friend in Banjul sent a telegram to her husband in Kingston, Jamaica, but, unfortunately, the message became hopelessly garbled. The man was interrupted during a business meeting by a secretary who informed him that his wife had taken seriously ill and he would have to take special blood back with him for her to be transfused. Needless to say, the poor man became panic-stricken. When he later discovered that the blood was "only for a chimpanzee," he became incensed and refused to cooperate.

Meanwhile, the staff at LEMSIP was trying to find a way to deliver the blood and medical supplies safely to the businessman at Kennedy Airport. This was where Fred Davis, the associate director and business manager at LEMSIP, came in. Fred was one of those people who seemed always by some miracle to know someone, somewhere, who could be relied upon to help out in any type of emergency situation, whether it was to get last-minute tickets to some ball game, or find the best specialist in New York City for someone in medical need. Fred was a "bean counter" with a big heart.

Fred knew an official at Pan Am and set about convincing the man of the importance and humanitarian need of getting Lucy's blood to Africa. "The blood has to be kept at the right temperature," Fred insisted to the airline official. "It can't go in the hold, or it might get too cold and freeze." A plan was agreed upon, and Fred, with the box containing the blood and the necessary medical supplies, met his Pan Am contact at Kennedy.

Lying in wait like two hawks, Fred and the Pan Am official descended upon the unsuspecting businessman as he disembarked from the airplane from Kingston. He was not amused, and despite their imploring, he steadfastly refused to have anything to do with the box of blood. "Besides," he said, "I have an important business meeting at the airport with some associates."

"No problem," retorted the Pan Am official, whereupon he offered the man and his colleagues the quiet seclusion of the meeting rooms of one of the private clubs at Kennedy, all to be wined and dined by Pan Am, free of charge, provided the man agreed to deposit his own luggage in the hold, and, in its place, take Lucy's blood on board with him to London. The poor man finally caved in, only to find the same bullying treatment when he got to London airport from officials of Caledonia Airways, the company whose flight continued on to Banjul. Fred and the Pan Am official had seen to that, too.

Janis and I, with no way of knowing whether our plans had come to fruition, drove back down to Banjul late on Saturday night so that we would be on time for the arrival of the London flight at five o'clock the next morning. We waited in the darkness on the edge of the tarmac of the little airport for the plane to arrive. Straining our eyes in the dim predawn light, we watched each passenger descend the stairway from the plane. After the fifth or sixth passenger had emerged, we saw a man clambering down the steps, carrying a very large cardboard box in his outstretched arms. This must be him, we realized. We raced to the exit doors of the arrivals lounge to wait for him to clear customs. Twenty minutes later he appeared, and as he made his way toward us, we could see that he had the most angry expression on his face. He walked straight up to me and, without a single word, jammed the box into my arms and continued on his way.

Relieved and delighted, Janis and I jumped in the Land Rover and raced back upriver to transfuse Lucy. It took two transfusions over a period of three days to bring Lucy's hematocrit close to normal. Everything turned out successfully, and she made a remarkable recovery. I have often wondered over the years whether the businessman ever came to realize what a wonderful deed he had performed.

Chapter Seven

■ ■ ■

WHERE DO I BEGIN? I asked myself. Marie-Paule and I agreed that if Molly's little puppy had any hope of surviving, I would have to find some way of giving her a blood transfusion. Eradicating her fleas and getting her on an easily digested diet were important, but they were only short-term remedies.

If I can't find a veterinarian, I thought, maybe I should try to locate local doctors' offices. Perhaps one of them would give me the equipment I needed to perform the transfusion myself. I again consulted the yellow pages. There were three physicians listed for Black River, the nearest town, less than an hour's drive west along the coast from Treasure Beach, and four doctors for Santa Cruz, a little town situated about fifteen miles to the north at the base of the steep Santa Cruz Mountains, reachable only by narrow, switchback roads.

I had already noticed black thunderclouds massing over the mountains even before noon, promising the usual afternoon rainstorm, with the likelihood of flash flooding. Within seconds, the mountain roads can be turned into raging torrents, the visibility reduced to nil by the thrashing rain. It can be quite a frightening experience, and the last thing I wanted was to get stuck in such a storm. It wasn't only the rain I feared; if I was delayed in my quest, I might also have to risk driving after dark. With no city lights to illuminate the sky, Jamaican nights can be every bit as black as a moonless African night. They make you feel as though you've been wrapped in a thick blanket, cut off from all sensory perception.

I decided, therefore, to try Black River first. Only if I had no luck there would I attempt to make my way back through the mountains to Santa Cruz.

I made a list of the doctors' names and addresses in each town, and then set about writing out a shopping list of the various supplies I would need. I would need syringes, a minimum of two of 35 milliliter and two of 60 milliliter capacity, in case I ran into trouble and had to repeat the procedures. I would also require an array of hypodermic needles: perhaps 25 gauge, 23 gauge, and some larger 21s, depending on the size of the puppy's and her mother's veins. It would be good to have some butterfly catheters, which would give me a lot more maneuverability if my patients decided to squirm. A bottle of heparin, an anticoagulant to stop the blood from clotting in the syringe, was absolutely essential, and it would be judicious to have a vial of vitamin K, the antidote to excess heparin, which I might need in case the puppy started to hemorrhage from the countless flea bites. It would also be good to have a blood filter in case some microscopic clots developed in the mother's blood as I drew it. Any clots would certainly kill the puppy if I inadvertently injected them into her, causing embolisms in her lungs or brain. Finally, I realized I'd need sterile gauze, some alcohol wipes to sterilize the skin, and a good slug of broad-spectrum antibiotic to treat the puppy's infected eyes.

I hurriedly took a shave and shower and, not wanting to look like a slob, polished my brown leather shoes, put on a freshly pressed pair of trousers and a shirt, and, in spite of the heat, a tie. Jamaicans can be very proper, in an old-fashioned way, when it comes to special occasions, and I didn't want to appear out of place. Marie-Paule packed me a cooler with a bottle of Red Stripe beer and a couple of bananas, in case my journey turned into an odyssey, and, in a less than optimistic frame of mind, I set off for Black River.

This drive, except for the rickety barbed-wire fencing along the side of the road, always reminded me of the open scrublands of Africa. Thorny acacia trees, grown lopsided from the constant blow of the trade winds, spotted the countryside, and huge zebu and Brahman cattle, like wildebeest, grazed the coarse grasses.

I entered the little town of Black River over the metal bridge that spans the estuary and parked the car by the fishing port. Immediately I was set upon by a throng of young men offering the best rate on the black market for Jamaican dollars in exchange for American money.

It was already two o'clock, and I could feel sticky patches of sweat developing under the arms of my dress shirt. I was on Main Street, where the three doctors' offices were located, but I had no idea exactly where they were. Most of the buildings were two-story wooden structures, with an occasional concrete monstrosity wedged in between. The upper stories of the wooden buildings, which protruded out over the irregular wooden boardwalk, provided welcome protection from both rain and scorching sun, their rusted corrugated tin roofs a broad array of browns and red. Everything was hustle and bustle, yet in the wonderfully relaxing slow motion of the Caribbean. The sidewalks were crammed with people: elderly fat ladies in floral print dresses; pretty young girls in the latest fashions, their forearms raised and wrists bent gracefully, handbags swinging from their elbows, promenading as if on the Champs Elysées. Old men stood in groups bantering in patois, and young blades lolled against the supports of the arcade, swaying from side to side to the rhythm of loud reggae music blaring from their shoulder-hoisted radios as they eyed the young girls passing by. Goats, a large black sow, and swarms of laughing children in uniform returning from school tried to make their way through what little space was left to walk, everyone spilling out onto the narrow roadway, competing for room with the belabored cars and smoke-belching lorries. Smartly uniformed policemen valiantly tried to maintain a semblance of order. A little way down a side street, at the edge of the sprawling market, was a one-story, yellow wooden hut—the local barbershop. Children sat sideways atop bicycles propped against the wall, gawking through the frameless open window at the three barbers inside busily cutting the hair of towel-shrouded customers, everyone—but everyone—swaying in synchronized rhythm to the catchy music from a radio, oblivious to the dangers of the sharp razors and pointed scissors in full swing. The whole scene reminded me of a Broadway musical.

Further up the street I glimpsed the sign for a pharmacy. That's an idea, I thought; perhaps I could get what I need there. In my trips to West Africa I had found that often you can buy just about any drug you want, from antibiotics to Valium, if you know where to look and how to ask; maybe, just maybe, Jamaica would be the same. At the very least, I could inquire as to the location of the three doctors' offices on my list.

Three elderly ladies selling fruit and vegetables sat on their haunches against the wall. I cautiously threaded my way through the piles of produce stacked neatly on the boardwalk and entered the pharmacy through the polished wooden door. What a difference from the bustling street. Everything inside was quiet, with an atmosphere of cathedral-like sanctity. The shop was lined with heavily burnished wooden shelves and glass counters bearing old-fashioned glass flasks and apothecary jars of brightly colored liquids, bringing back a flood of childhood memories of the chemists' shops that existed in ancient English and Irish villages in the time before pharmacies had to compete with supermarkets for the sale of panty hose, cigarettes, trashy magazines, and condoms. Three sedate little old ladies in colored straw hats sat against the wall in a row of wicker-backed chairs, their lace-gloved hands folded on their laps, presumably waiting for their prescriptions to be filled. "Do you sell hypodermic syringes and needles?" I inquired from the lady behind the counter. Her eyebrows immediately arched as she looked up at me over her prim half glasses. She thinks I'm a junkie trying to buy drug paraphernalia, I realized.

"We sell syringes only for diabetics," she replied sanctimoniously, stressing the word *only,* "and only then by prescription." The little old ladies looked up at me reproachfully.

Suddenly feeling very out of place, I asked, "Could you please tell me where Dr. Brown's office is located?"

"About six chains farther up the street, on the right, past the church," she replied curtly. Nothing was turning out to be easy. She was giving me directions based on the ancient Anglo-Saxon system of measuring distance in chains and links, which I had learned as a child but had all but forgotten till now. I had made spectacle enough of myself in front of these august ladies, and I decided it was better not to belabor the point by asking for more specific directions.

Out on the busy street once again, I made my way past the church, all the while scanning every sign and billboard for a clue to Dr. Brown's. As I approached a stately old house with sagging verandahs, peeling white paint, and overgrown bougainvilleas, I spotted a rickety wooden sign, swaying in the wind, bearing the faded words DR. BROWN, PHYSICIAN. I began to rehearse in my mind exactly what I would say. "Good afternoon, doctor," I would say. "My name is Doctor James Mahoney, from the New York University School of Medicine, here on holiday." Would that impress him? Probably not. Another approach might be more appropriate: "I have a problem," I would humbly say. "Perhaps you would be kind enough to help me." Jesus, I thought, that won't work either. Let's start again. "Good evening, Dr. Brown, a friend told me that you might be the very person to help me with a special problem. . . ." How do you approach a doctor you've never met before and tell him you have a three-and-a-half-week-old runt puppy who needs a blood transfusion?

I climbed the verandah steps to find a woman waxing the already highly polished hardwood floor of what appeared to be the waiting room. As I stepped forward to speak to her, I slipped, falling heavily on my rump. Damn, shit, and piss, I said to myself, this is turning out to be a great beginning. "Is the doctor in?" I inquired meekly, trying to muster a false nonchalance as I picked myself up off the floor, the growing pain in my coccyx almost taking my breath away.

"No," the woman replied. " He won't be back until this evening." Well, that was the end of that. One down, two to go. I thanked the lady and, feeling an utter fool, I left.

I continued along the road out of town, which bordered the broad Black River Bay, all the while reading every wayside sign and billboard in search of the next doctor on my list. About a quarter of a mile farther on, I found the far less pretentious office of the second doctor on my list. It was a whitewashed concrete-block affair. A young lady there quickly informed me that the doctor wasn't in and wouldn't be back until later that night. And so I made my way to the last doctor on my list for Black River. I was now about half a mile out of town. The house was a large, sprawling bungalow, and I had difficulty finding the entrance.

I finally found the doorbell and the woman who answered the door politely informed me "The doctor is not in, and will not be back until tonight."

"It's amazing," I said, frustrated. "I've called on all three doctors in Black River, and not one of them is at home. I wonder where they might be?"

"Oh, they all go fishing upriver on Wednesdays," she answered. I would have found this funny had I not been so desperate. I explained to the lady my problem, and she suggested I go to the hospital, a short distance farther up the road.

It was well into the afternoon by now, and I wondered whether I would find what I needed before nightfall. The hospital, its white-painted walls also peeling, was set back in a large fenced area. Outside the fence, a roadside sign proclaimed SILENCE ZONE, but inside all was anything but silence, with the bustle and hum of vendors selling beer, fruit, cigarettes, and jerk pork, catering to the large crowd of visitors, hospital employees, and ambulatory patients who moved dejectedly through the crowd in their faded blue bed gowns. The lower level of the two-story building was the men's ward; most of the mosquito screens on the tall, narrow windows were either torn or missing altogether. To the left of the main building was a smaller, single-story structure of white-painted concrete block. The sign above the door indicated it to be the outpatient department.

I was overwhelmed by what I saw as I entered the concrete building. Everything seemed to be in motion—patients in dressing gowns wandered around aimlessly or propelled themselves in wheelchairs, groups of orderlies and nurses chatted and laughed with one another or with relatives and friends of patients, everyone coming and going in busy confusion. The main part of the waiting room was filled with high-backed wooden benches arranged in rows, like church pews. The walls were off white above and battleship gray below; a broad irregular band of greasy black dirt, created by the rubbing of countless hands and heads of patients waiting for attention, separated the two. Broad louvered windows on either side of the waiting room allowed a cool breeze to carry away the heat inside. High on the back wall was a sign that read EMERGENCIES ONLY ON SATURDAYS AND SUNDAYS, the precise meaning of which I found difficult to comprehend.

I studied the churning mass of people, trying to determine who best to approach for help. In the distance, I saw a friendly-faced lady, smartly dressed in a light brown uniform, obviously a person of some authority. I approached her and began to explain my story. She pointed off into the crowd. "You had best speak to the matron, over there," she said, as she picked up the mop that she had left leaning against the wall and resumed washing the floor.

After some moments, I spied what I knew for sure to be

the matron. In her blue uniform and starched crimpled white cap, she had the same unmistakable authority that I had come to know from matrons in the large hospitals of London. When the matron refers to Mr. Smith, you know by her very tone of voice that she means Mr. Smith, the eminent surgeon (surgeons are titled mister, not doctor, in Britain), and not Mr. Smith, the janitor. Everyone, from the lowliest student nurse to the top surgeon, stands in awe of this dowager. The phenomenon of the hospital matron is quite nonexistent in the United States.

For what seemed like the hundredth time, I began to explain my tale about the puppy and my need to give her a blood transfusion. It was difficult to decipher her facial expression and tell whether she had any sympathy for my plight, but eventually she pointed to a row of chairs at the very back of the building and ordered me to sit while she consulted with the doctor on duty.

She disappeared through one of the side doors in the room, the doorknob of which was missing. Five or ten minutes later, she reappeared. "The doctor will see you in a moment," she curtly informed me as she spun on her heels and disappeared once again into the crowd. I settled down to wait. Marie-Paule's beer would have gone down well now, I thought as I tried to lick my parched lips and swallow, but I had left the cooler in the car. Just as well; beery breath wouldn't be such a great idea when I had to speak to the doctor.

I looked around at the wide variety of waiting patients. A mother held her little girl, the daughter's face pressed against her mother's chest; a doleful old man sat with one side of his swollen face wrapped in a broad bandage, probably due to a tooth abscess; and a younger man walked endlessly around in aimless circles, a deep blood-caked gash extending from his left cheek down the side of his throat to his chest.

Seeing the misery before my eyes, I began to wonder what right I had to bother a doctor when he had all these problems to contend with. This was Jamaica, after all; perhaps not rife with the poverty, malnutrition, and rampant disease you might find in Africa, but here the concern for an animal would have to come second to that for a human being. I considered walking away and trying to think of some other way to solve my problem. But then again, life is life, I thought, whether it be human or nonhuman. All I was asking for was a little help; I would do the work and pay for the materials. Convincing the doctor would be the real test of my Blarney, an Irishman's power of persuasion.

Every time a patient entered or emerged from the examination room, I jumped up from my seat to see if I could catch a glimpse of what was going on inside, or maybe even attract the attention of the doctor on duty—a well-timed cough might be a diplomatic way to remind him I was still outside. But my efforts were to no avail. Patients came and went, except for the man with the gashed throat, and I began to think that either he and I had been forgotten or the doctor was determined to make me wait until the bitter end. By now, I had been waiting for two and a half hours. For the very first time in Jamaica, I wondered whether my being white put me at a disadvantage. Yet that was grossly unjust of me; there were people here whose needs far exceeded mine.

The man with the gash in his throat was beginning to worry me. He had been there when I first arrived at the hospital, and a lot of people had come and gone since, their medical problems taken care of. I walked over to him and tried to strike up a conversation. His accent and patois were extremely difficult for me to understand, and, to top it all, he had a tendency to stutter. He was a lorry driver and had got entangled in a brawl in one of the mountain villages, where another man had attacked him with a machete. The police had been called, the attacker arrested, and the wounded man brought to the hospital. His wound was no longer bleeding profusely—it was now caked in a dirty brown sludge—but it still looked very serious. I took the liberty of searching out the matron I had spoken to earlier. I eventually found her and began to express my concerns about the man. It turned out that the police had inadvertently given the name of the attacker—not of the injured man—to the hospital staff. Consequently, every time the nurse came to the door of the examination room, she called out the wrong name.

This was fortunate for me, as it turned out. Not only had I helped the lorry driver, but the doctor, who was inside the examination room, overheard the commotion in the hallway and peered out the door. Upon catching sight of me, he grasped his cheeks with both hands and, with the broadest, sweetest smile, explained that he had totally forgotten I was waiting. His name was Dr. Francis. He had a round face and silvery hair and reminded me an awful lot of an older version of James Earl Jones. He had long since retired from active hospital duty, but he volunteered here two afternoons a week to help out with emergencies. I explained that I was on vacation in Jamaica and came from the New York University School of Medicine. I don't think he could have cared less if I had come from Mars; once he found out about the puppy he was willing to help me, and that was all that mattered.

Dr. Francis guided me gently into the examination room. It was a large, shabby room with a high ceiling and faded white walls. Three Formica-covered desks were strewn with paperwork, gauze, cotton balls, thermometers, and stethoscopes. Two young, white-coated doctors sat, each to a desk, examining patients. Only later did I discover from one of these young doctors, as he took a welcome cigarette break, that he and the other young doctor were students, carrying out their internships from Blackfriars Hospital in London. "Cor blimey!" he said in typical Cockney expression, between deep puffs, "it's like being thrown in at the deep end here."

I could well appreciate his frustration. I doubted that the lectures he attended or the medical textbooks he studied prepared him for this scene of raw, basic medicine in a developing country. There would be no CAT scans here, no MRIs, nor any other advanced diagnostic tests at his disposal, no bevy of medical specialists with whom he could consult. In Jamaica he would have to learn to live by his wits, apply common sense, and, above all, develop his sense of compassion. Yet I knew he was learning something far more significant than his classmates who had chosen the comfort and order of some prestigious hospital in the United Kingdom to further their medical training.

Dr. Francis took my shopping list and, holding it at arm's length, his spectacles perched on the end of his nose, he began to strike off each item he couldn't supply. "We don't have that," he mumbled, shaking his head, "or that . . ." Gone were the vitamin K, the 60-ml syringes, most of the hypodermic needles, the catheters and the blood filter, and, worst of all, the antibiotics. But he did have one 35-ml syringe, one 12-ml syringe, two 21-gauge needles, a single alcohol swab, and, most important, a small vial of heparin.

These were the bare essentials, but I had what I needed to draw the blood from Molly and transfuse her puppy.

Dr. Francis briefly asked me about my work, and I told him that I was involved in AIDS research, to which he said, "Please find a cure soon." With that, he gave me a warm handshake, and we said our good-byes.

Chapter Eight

■ ■ ■

As I MADE MY way along Main Street, I couldn't help thinking of the young medical student I had just met. His anguished expression and the way he puffed nervously on his cigarette were clear signs of the stress he was under. I could identify with that. I remembered my own feelings of inadequacy and the overwhelming fear of being out of my depth when, at around his age, I first started working with primates at the Royal College of Surgeons of England, in 1967. I was certainly thrown in at the deep end as I took my first tentative steps into a life of research, to study the reproductive physiology of monkeys, animals about which, for all practical purposes, I knew nothing.

Oddly enough, reproductive physiology had been my worst subject in a generally poor showing as a veterinary student in Glasgow—"one of the worst students to graduate since the war," as Professor Sir William Weipers, the dean of the veterinary college, unceremoniously informed me on the day of graduation. Yet, after I had been in practice in Scotland for a year or so, I began to realize that as much as 80 percent of what I dealt with as a farm vet depended, in one way or another, on a deep knowledge of reproductive physiology. I decided therefore to enroll in a master's degree program at the University College of Wales in Bangor, North Wales, to study "comparative mammalian reproduction and embryology." I felt sure that with the knowledge I'd gain, I could return to practice a much better vet than I had left it.

This was the most carefree stage of our married life, a brief respite before the coming responsibilities of family and career. Marie-Paule and I lived in the charming little village of Beaumaris, on the island of Anglesey, one of the sunniest spots, by some quirk of geography, in the whole of the British Isles. Its twelfth-century castle, built by the Normans, was surrounded by a moat to keep the wild Celts at bay. Our apartment had a broad bay window that overlooked the Manai Straits, the inlet off the Irish Sea that separates the island from the northwest coast of the mainland. We had a breathtaking view of the rugged mountains of Snowdonia and were awakened each morning by the soothing sound of the sea lapping against the pebble beach, and the cries of the seagulls circling overhead. Here, everyone spoke Welsh, a daily reminder that the English had never quite succeeded in their centuries-long conquest of Wales. With its singsong lilt and impossible mixture of ch sounds (as in the Scots word *loch*) and double L's (a cross between the th of *thick,* and the liquidy l-sound you might make as you are about to produce a juicy spit), it was the first language Pádraig, our first child, heard as he made his bumpy debut into the world with the help of two Welsh nurses.

The program I entered offered the only master's degree

of its kind in the whole of Europe in those days, and the four of us accepted were instructed by twelve professors. I soon discovered, for the first time in my life, the sheer joy of learning for learning's sake. We studied the mysteries of reproduction and breeding of elephants and shrews, kangaroos and wombats, yaks, cats, chickens, and just about all the rest of God's creation, except, amazingly enough, primates.

One of the courses that the four of us had to take was histochemistry, the study of techniques for staining tissues and cells and then examining their features under a microscope. From their chemical characteristics, we were to determine their physiological functions. The tissue we were given to study over the course of the academic term was from the uteri of pregnant mice. We were required to embed the tissues in a special wax, cut them into thin sections using an instrument called a microtome, and mount them on glass slides. We then subjected them to the various stains that had been provided to us. After several eye-straining days of peering through the microscope at the wide variety of cells scattered throughout the tissue, I was surprised to notice large numbers of huge, foamy cells, each with a large indented nucleus and cytoplasm packed with purple granules. These cells formed thick, multilayered cuffs around the small arteries in the uterine wall that led to the placenta. The cells were spectacularly beautiful, but I didn't have the faintest idea what they were. My professors were of no help whatsoever; each one had a different opinion as to what the cells might be. This is incredible, I thought. Maybe I've made a major discovery. If this is research, then give me more!

I quickly set about drawing up a plan of how I might study these cells in order to discover what they were, and what they did. There was no doubt in my mind that they served a very important function, judging from their sheer numbers. At what point in pregnancy did they first appear? What were the granules made of? What happened to the cells at the end of pregnancy? I was consumed with excitement and a sense of adventure, but there was one thing I hadn't taken into consideration. These tissues, which had been prepared and preserved by someone unknown to me, had come from living, breathing animals, not just from the glass jar in which I had been given them. To carry out my ambitious plan of research would require my selecting a dozen or so female mice, having them bred, and then killing them at different stages of pregnancy, so I could dissect their uteri for examination under the microscope. (Looking back on it, I realize how hopelessly naïve my estimate of a dozen mice was; it would have required far more animals than that.) The killing would be painless, I assured myself—just a brief exposure to a cotton wad soaked in ether, and the mice would fall gently into a deepening sleep. But it wasn't like that at all. The mice, little white ones with pink noses, rubbed their faces and eyes with their front paws in frenzy, trying to rid themselves of the noxious fumes, their long whiskers twitching frantically. I tried not to look, to turn my back after I had dropped each one into the ether jar, but it didn't work. I couldn't avert my eyes—I was killing creatures with a sentience all their own. "If this is what it takes to find out about these bloody cells," I said to myself, "then to hell with it," not exactly the attitude one would expect from a budding young scientist.

As it turned out, I was spared the need to sacrifice any more mice. One afternoon as I sat at the lab bench, thumbing through the pages of a recent edition of *The Journal of Anatomy,* my eye was suddenly arrested by a large photograph. It was a picture of my cells. There was no doubt about it; these were the same cells that I had been studying day after day, long into the night, for the last several weeks, the memory of their characteristic architecture now etched deeply into my mind. The cells even had a name, I discovered—metrial gland cells—and had first been described back in the early 1930s. One by one, I crossed off the experiments I had written down on my list, all but one of which had already been carried out by others, long, long before me. I was devastated. I felt like the milkmaid in La Fontaine's fable, who, on her way to market, dreams about all the things she is going to buy after she sells the pitcher of milk she is carrying on her head. Needless to say, she slips and breaks the pitcher. This was not the last pitcher of milk I was going to spill in my scientific career, but it was certainly the most poignant. Ironically, to this day, thirty years later, scientists still have not figured out what functions these cells serve or what is the chemical composition of their granules.

It was then I knew I was hooked; I had fallen in love with research. I decided to carry on, but in a field that didn't require the killing of animals. I suppose, looking back on it, I didn't have the courage to face this issue at the time, wanting my cake and eat it, too.

One problem that had fascinated me as a vet, and which has major implications in farming, was retained placenta, a condition that commonly occurs in cows after they have calved. The placenta doesn't separate from the womb as it should within the first few hours after the birthing, and it soon becomes contaminated with bacteria and a life-threatening infection of the uterus sets in. Even after successful treatment with antibiotics, the cow is likely to be rendered infertile.

Intent on studying this disease, I enrolled for a Ph.D. at the Royal College of Veterinary Surgeons in London—the same veterinary college that had turned me down several years earlier because I hadn't seen a chicken lay an egg—to study under a very famous reproductive physiologist, Professor Amoroso. Amo, as students and colleagues affectionately referred to him, was half Spanish, half Jamaican, and the first black person ever to be accepted into the Royal Society, the most prestigious scientific body in the world.

Marie-Paule and I were penniless by then, having used up all our savings to support us while I was studying for my master's degree. If it hadn't been for my wife meeting, by sheer chance, an American nun, we would have been in a terrible jam. The young nun, Sister Marie-Joseph, introduced Marie-Paule to the mother superior of her convent, which belonged to a French order, and the nuns hastily made arrangements to convert the top floor of the convent into an apartment for us, at a very modest rent, when we could afford to pay. I was the first man permitted to sleep in any of the convents of the order in its several-hundred-year history. Despite the nuns' help, with Pádraig barely more than a year old and Nathalie well on the way, Professor Amoroso explained that there was no way we could make it as a family in London on the meager stipend I had been awarded from the Scientific Research Council. With an impish chuckle, Amo told me not to be discouraged and pulled a crumpled newspaper clipping from the back pocket of his trousers. It was an advertisement for a position as veterinarian in the dental department of the Royal College of Surgeons of England, for work with monkeys. The advertisement didn't specify exactly what the job entailed, but I was well aware that I knew nothing about monkeys and didn't really have an interest in dental research. This job wasn't for me, I informed Professor Amoroso. "But they don't want someone for dental research," he said. "They're after a vet to care for the monkeys and at the same time study their reproductive physiology. The field is wide open and full of opportunities. At least look into it," he encouraged me as he handed me the bus fare to go straight away for an interview.

The Royal College of Surgeons, with its rather awesome granite steps and ornate colonnades, is located in Lincoln's Inn Fields, within the heart of the ancient square-mile City of London. It purportedly had housed Napoleon Bonaparte's liver in its Hunterian Museum, until a direct hit during a bombing raid in the Second World War destroyed it.

After an hour-long interview, the professor of the dental department drove me the twenty-five miles through London and out into the Kent countryside, the Garden of England as it is known, to visit the primate facility. The laboratory was tucked away along a narrow, winding lane, amid apple orchards and rolling green hills, on the edge of the tiny, picturesque village of Downe, the village where Charles Darwin had lived.

After dressing me up in protective outer clothing—disposable rubber gloves, surgical mask, hat, white coat and booties, designed as much to ensure the animals didn't contract human disease as to protect people from their microbes—the professor took me to see the monkeys. Crab-eating, cynomolgus, or Java macaques, as they are variously called, have pixie-like pointed ears, a strikingly pointed coif atop their heads, and long tails. They were the very picture of devilment. I had never before come face to face with an animal that showed such inquisitiveness, and the rambunctious way they rattled the bars of the cages at me, a total stranger, captivated my imagination.

To my total amazement, the little voice inside me urged me on, and I found myself telling the professor yes, I would take the position. Yet, I could never have imagined starting a new job in a more abject state of ignorance. Like the medical intern I had met at the hospital in Black River, I had an overwhelming sense of being out of my depth. Apart from knowing that monkeys—at least Old World monkeys—menstruate like women, and are more intelligent than most animals, I had no knowledge or understanding of their physiology or the diseases that affected them. I stared at the macaques in their cages, wondering how I would ever set about taking one's temperature, or putting a stethoscope on its chest to listen to its lungs, without having my fingers bitten off. I would have to start from scratch before I could even begin to contemplate being an effective primate veterinarian. And I hadn't the foggiest notion how to go about doing research on them. Yet I knew it was finally good-bye to cows and placentas, and all the other things I had known.

My first day in my new life of primate research didn't start out well. I was late arriving at the lab, after getting hopelessly lost in the Kentish countryside. When I did arrive, around ten-thirty that morning, I found the director of the lab awaiting me with an anesthetized monkey to examine and two young women technicians in attendance to assist. I could tell by his face that he wasn't the least bit pleased by my tardiness, and I knew any excuses I might offer would do nothing to appease him. He handed me a syringe and needle and asked me to draw a blood sample from the monkey. She was lying on her back on the examination table, and I rolled her onto her abdomen to look for a vein beneath the upper surface of her forearm, the same vein I would use if she were a dog or a cat. With an outburst of impatience, the director elbowed me sharply to one side, grabbing the syringe and needle from my hand. Turning the monkey onto her back again, he inserted the hypodermic needle into her femoral vein, the blood vessel that runs deep down the inside of the thigh.

"Didn't they teach you anything in vet college?" he said, in an obvious pique of disgust. I began to worry what else might be in store for me.

The director next asked me to perform a rectal examination on the monkey, to determine whether she was pregnant. Turning her on her right side, I inserted the lubricated finger of my gloved right hand into the monkey's rectum, only to realize I had entered the wrong orifice. Oh, God, I thought to myself: first I'm late, then I make the mistake of choosing the wrong vein, now he's going to think I don't know the difference between a rectum and a vagina. I felt my face flush with embarrassment, and saw the two technicians quickly avert their gaze toward the ceiling no doubt saying to themselves, Cor, we've got a right one 'ere!

In those days, the late 1960s, medicine had yet to make the advances it has achieved today in the understanding of the human reproductive process. Radioimmunoassay, a now common technique for measuring hormone levels in the blood as a way of determining ovulation and other reproductive events, was still several years away. There were no simple pregnancy tests available, and laparoscopy, the surgical use of a thin telescope-like instrument to visualize the ovaries, had not yet been introduced into human medicine. Needless to say, when it came to monkeys, science was still in the Dark Ages; little was known about even the most basic aspects of their reproductive physiology or fetal development.

This lack of knowledge was a major block to research into the development of vaccines that might protect young children from future dental cavities. The researchers at Downe were particularly interested in using the monkeys as models to study human dental disease, and they hoped to determine how, for example, the oral cavity of the newborn, sterile at the moment of birth, becomes populated with bacteria, especially those bacteria that later cause cavities. For the researchers to carry out such studies, I would have to deliver infant monkeys by cesarean section under sterile conditions. But this required that I first come up with methods for determining ovulation, so I could ascertain the time of conception and from this calculate the length of pregnancy. I would next have to find ways of predicting when labor was about to occur, so that I could interrupt the pregnancy at the last moment and deliver the infants as close to term as possible. Above all, I would have to avoid natural labor from occurring; otherwise, the mothers would contaminate their infants' mouths with bacteria as they licked them dry of birth fluids during the first few moments after delivery. I had my work cut out for me, but at least I had a solid basis for a Ph.D.

As strange as it seems to me now, I didn't at the time see anything cruel or inhumane in the way the monkeys were kept. Their cages were scrupulously cleaned each day, and some of them were even permitted a certain degree of social life. The technicians who looked after them, mostly young men and women barely out of high school, showed a great deal of love and affection for the animals and never shortchanged them for attention. Even when the animals did develop cavities, which are incredibly difficult to induce in monkeys, and require their being fed very sugary diets, their teeth would be drilled and filled, under general anesthesia no less, using the most up-to-date dental techniques. As they grew older, the monkeys became candidates for the study of gum disease and other degenerative oral conditions of old age, which also commonly occur in human beings, but none was ever left to suffer without treatment.

Furthermore, my own research dealt with the wondrous, hopeful beginnings of life, not with death and suffering. I wasn't the least bit disturbed by the animals' lack of freedom, their confinement to cages for their entire lives. I would avoid these particular concerns—not by design, I might add, but through sheer good luck and fortune—for many years, through my year and a half in Oregon, where I would complete my postdoctoral training, and through the four years I would spend at the Primate Center in Madison, Wisconsin. The inevitable head-on collision would not occur, in fact, until I had been working at LEMSIP in New York for a few years—and all because of three special little animals I would get to know there. For the present, however, I was able to have all the satisfaction of research without the gnawing feelings of guilt and doubt about the way animals are used in research. Only slowly would my attitudes start to change, as I began to learn about the behaviors of the different species and recognize that each animal, in its own way, is uniquely different from the rest. Each has its own aspirations and sense of value of itself; each responds in its own way to the world around it; each has its own fears and trepidations. Years later, a scientist would say to me, "It must be fascinating to work with chimps; not like rhesus monkeys—they're so boring." How wrong he was. No animal is boring; it's only the monotonous conditions under which we keep them in the laboratory that give this misconception, depriving them, as we do, of any possibility of expressing their individuality.

In addition to being the largest and most beautifully situated of the seven primate centers funded by the National Institutes of Health, the Oregon Regional Primate Research Center boasted the largest number of species—ranging from tiny bush babies and other prosimians to a variety of macaques, including the Celebes macaque, or black ape, and the Japanese macaque, or snow monkey. The main thrust of research at the center was behavior and reproductive physiology, providing me ample opportunities to continue my own studies on ovulation and conception.

Looking back on it now, I suspect my attitudes did start to change when I arrived in Oregon in the summer of 1972. Yet, like so many medically oriented researchers, I had begun by thinking of behaviorists as little more than binocular-toting individuals obsessed with asking the same mundane and obvious question over and over again: "Why did the chicken cross the road?" My year and a half in Oregon opened my eyes to what behavioral research is all about, and taught me that to be a good medical researcher or veterinarian you'd better understand the behavior not only of the species you deal with, but also of each individual animal.

Instead of returning home to the British Isles in late 1973, as I had presumed my family and I would, I was offered a position at the Wisconsin Regional Primate Center, in Madison.

The director of the center, Dr. Robert Goy, had been at Madison for a little more than a year when he asked me to come and work for him. He wanted me to take charge of the surgery, reorganize it, and tighten the rules about who had access to it and under what conditions. Goy was fed up with the laissez-faire attitude of some of the scientists at the center who, after years of being free to do just about anything they pleased with the animals, were now resisting some of the changes Goy wanted to implement. No rule had existed, for example, that a veterinarian had to be present during surgeries, let alone be the one responsible for performing the procedures. There was no control over anesthesia, no one empowered to stop a surgical procedure if it got out of hand and the animal began to wake up.

In particular, Goy wanted to curb researchers who, just because they had a Ph.D. and were recognized experts in their fields, looked upon themselves as surgeons, having learned their surgical "expertise" from their mentors, who were usually no more than doctors of philosophy themselves.

At Madison, I worked with Steve Eisele, a shy man in his early thirties who was in charge of the center's rhesus monkey breeding program. Where some technicians work silently with the monkeys, Steve was all talk. Whether the monkeys actually understood what he was saying to them or just picked up on the intonation of his voice, I don't know, but he certainly had an extraordinary ability to communicate with them, and I could tell they trusted him implicitly. I immediately bonded with Steve, and we jumped into all sorts of fascinating research together.

Steve had devised a highly successful, though frustratingly arcane, method of telling when the monkeys were about to ovulate by studying the daily changes in intensity of the red coloration of their perineal sexskin, the area around their bottoms and base of the tail. Steve had developed another rare skill—the ability to obtain semen from the males. He had produced a couple of hundred rhesus monkey babies over the years by single artificial inseminations of females, a number many times higher than anyone else in the world has ever achieved. The combination of these two skills, the ability to tell when the female was about to ovulate and the competence to be able to obtain pregnancies through artificial insemination, presented all sorts of opportunities for interesting research. In particular, I had long been interested to determine how long sperm and ova retain the capacity for fertilization, a study that would be just about impossible to carry out in human beings.

Steve and I, together with another young researcher named Jerry Robinson, set about developing a research plan to determine how long an egg remains viable after it's released from the ovary. We would inseminate each female only once in each cycle, hoping that, in retrospect, we would find that we had carried out this procedure at varying times over a period ranging from seventy-two hours before to seventy-two hours after ovulation. Repeated in a sufficient number of monkeys over two or three menstrual cycles, we hoped to determine for how many hours the sperm and ova remained viable and capable of producing pregnancy. We would need to collect blood samples for hormone analysis throughout the cycles, concentrating our efforts on sampling every six hours around the time we expected ovulation to occur.

When it came my turn to take the 3:00 A.M. blood samples, almost invariably one of the monkeys would escape from me and proceed to run around the room, jumping from one cage to the next, defying every attempt I made with kind words or pieces of apple to talk or cajole her back into her cage. After thirty minutes of failed effort the first time this happened to me, I realized there was no other way but to call Steve to the lab to help me get her back in her cage. "Steve," I said into the phone, "sorry to call you at this time of the morning"—it was by then about 3:45—"but . . ." I went on to explain in apologetic terms how the monkey had escaped me, and that even between the two of us, I thought it would be very difficult to get her back into her cage.

When Steve arrived, he walked silently past me and proceeded into the animal room, barely able to hide the grin of condescension on his face, and said to the monkey, "What is going on here? You just get right back in your cage, and let's have no more of this!" There was no chasing the monkey, no getting angry and yelling at her—she just walked right back into her cage. I was flabbergasted. The whole operation had taken him less than a minute. In truth, I think it was a conspiracy hatched between Steve and her, and all the other monkeys who would escape me over the coming weeks of the study.

Chapter Nine

■ ■ ■

As I drove the winding roads back to Treasure Beach, the transfusion supplies from Dr. Francis safely tucked on the seat beside me, I began to worry that I might not be able to find Molly before twilight. She might be off at Jake's or roaming farther afield. Without adequate light, I wouldn't be able to see to draw Molly's blood and transfuse the puppy.

On the way to find Molly I stopped by the villa to pick up Marie-Paule so that she could help me draw the blood. I also collected the scissors I would need to clip the hair on Molly's foreleg. With the precious few materials that Dr. Francis had given me, we set off for Miss June's house down the road. I was lucky. Molly was there to greet us, along with her other puppies.

I explained to Miss June and Miss Maisy what I was about to do, and using the little table outside the house as a work area, I laid the materials out neatly. First I swabbed the rubber stopper of the heparin vial with the one and only alcohol wipe I had, carefully returning it to its foil wrapper for later use. I drew a little of the heparin into the 35-ml syringe, enough to wet the inside of the barrel, and ejected all but one milliliter or so into the air, the volume I would need to insure that Molly's blood didn't clot once I drew it.

Then I put together a rough-and-ready tourniquet, made out of a broad rubber band and a ballpoint pen, and I wrapped it around Molly's foreleg, just above her elbow. Marie-Paule crouched low behind Molly, to prevent her from backing away from me as I withdrew the blood, her right hand holding on to Molly's right shoulder, her left hand twisting and tightening the makeshift tourniquet. Molly's brachial vein stood out clearly, and even in the fading light it appeared the size of a drainpipe. I reached for the scissors on the low table, clipped the hair over Molly's bulging vein, gave the skin a wipe with the alcohol pad, returning it once again to its aluminum foil sleeve, and prayed that I wouldn't screw things up; there would be no chance for a second shot.

"Good girl, Molly! Good girl!" I repeated over and over, through clenched teeth. "She's such a good a girl, this Molly," I continued, trying to soothe her. Holding Molly's leg with my left hand, I pierced the skin of her foreleg with the hypodermic needle, directly over the distended vein, and immediately felt the warmth of the dark, purple blood as it spurted into the barrel of the plastic syringe. Molly was as good as gold, and except for a slight jerk as the needle first pricked her, she sat calmly, licking my face, or turning her attentions to her foreleg, licking the site where I had inserted the needle, making it difficult for me to see what I was doing. Within a minute or so, the syringe was full, and I withdrew the needle, applying my thumb over the site to stop the bleeding. The tension over, I exhaled my pent-up breath. Molly had been a perfect patient.

Back to the villa we drove, the syringe full of blood tucked carefully under my arm to keep it warm. It was just about dark, and when we arrived at the villa I gathered as many table lamps as I could to give enough light to carry out the procedure on the puppy.

Marie-Paule sat in a chair close to the living-room table, the little puppy straddled motionless across her lap. I knelt down in front of her and took hold of one of the puppy's front legs, trying to feel for a vein. It was there, but very small. The 21-gauge needle that I had acquired from Dr. Francis seemed huge in comparison, and I knew there was absolutely no way I would be able to insert it into such a tiny vein. The pale purple veins that meandered across the wall of her bloated abdomen appeared larger. I knew it would be a mistake to even try these, however; even if I were lucky enough to thread the needle into one of them, I wouldn't be able to inject the relatively large volume of blood before the thin wall of the vessel ballooned and burst from the pressure.

I sat pondering the situation. I was worried that I had no vitamin K to act as an antidote to the heparin. Any excess heparin could cause uncontrollable hemorrhaging of internal organs, and even of the skin from the countless scores of flea bites the puppy had all over her. I also couldn't allay my fear that the blood might have tiny clots in it. This would be the puppy's death knell. I rocked the syringe back and forth very slowly as I held it up to the light, to see if I could detect any jellylike clots clinging to the side of the syringe. But even if I couldn't see any, it didn't mean there weren't microscopic clots. Without a blood filter, I wouldn't be able to trap them. As small as they might be, such tiny clots would be carried in the puppy's venous system to her lungs, clogging the delicate capillaries in the walls of the alveoli. Great cone-shaped segments of her lungs would be instantly deprived of blood. There would be no saving her then; she would die from necrotizing pneumonia. But this particular concern was academic, in any case, because I didn't have a needle small enough to give the blood intravenously. Maybe God meant it to be this way, I mused. It's funny how you sometimes find yourself talking to God, even when you're not sure you believe in Him.

There was no other way; I would have to give the blood intraperitoneally—into the space beneath the body wall that surrounds the intestines and other abdominal organs. At least I wouldn't have to worry about the dangers of clots in the blood, and the heparin would probably pose less of a risk, given that way. I have always found it hard to believe that relatively large cells, like red blood corpuscles, can actually make their way through the thin peritoneal lining of the abdominal cavity into the bloodstream. Yet they can, albeit very slowly. One other benefit of this route of transfusion over the standard intravenous method was that I wouldn't have to worry about the sudden overloading of the puppy's vascular system if I gave too much blood or injected it too fast. This procedure wouldn't be entirely without risk, however. If I inadvertently punctured the wall of her intestine as I passed the needle through the abdominal wall she would probably develop peritonitis, a potentially fatal abdominal infection.

Rolling the pup over onto her back, I clipped the hair from the lower right side of her tummy and cleansed the skin with the well-worn and almost dry alcohol pad. Slowly I began to insert the needle into the skin. In an effort of supreme concentration, my eyes tightly closed, my breath held firm, I began to feel my way, layer by layer, first through the skin, then the two muscle coats of the abdominal wall, until finally I felt the needle sigh as it pierced the peritoneum, the last layer before the abdominal cavity. I attached the small 12-ml syringe, which now contained dextrose solution, and injected slowly. The fluid went in easily, and the skin didn't swell around the injection site, telling me I was truly in the pup's abdominal cavity and not just under the skin or in the muscle layers. With extreme care, so as not to hurt her, I unscrewed the syringe of dextrose from the needle still inside her abdomen and attached the larger 35-ml syringe that contained the blood. Slowly I began to inject. The pup had yelped only when I had made the first puncture through her abdominal skin. After that, she lay perfectly still until I had finished.

"Well," I said to Marie-Paule as I exhaled and withdrew the needle from the puppy's abdomen, "it isn't the greatest, but it's the best I can do under the circumstances." I had no idea how long it would take for Molly's red corpuscles to be absorbed into the puppy's system; at least many hours, I thought, and probably several days. Given this way, the transfused blood wouldn't give the puppy the sudden curative jolt I was hoping for, but it would give her the boost she so needed.

As I watched the little puppy now lying fast asleep across Marie-Paule's lap, I thought about the primitiveness of my transfusion technique and the less than ideal conditions under which I had performed the procedure. It was crude, to say the least, compared with the highly controlled conditions that I was used to working under in the laboratory, handling dangerous blood-borne viruses such as HIV.

Chapter Ten

■ ■ ■

I REMEMBERED THE VERY first time I gave an intravenous injection of HIV, the virus that causes AIDS, to a chimpanzee. This was the beginning of LEMSIP's collaborative program in AIDS research with the Pasteur Institute in Paris, and Professor Marc Girard, the head of the program, had flown over from France in the Concorde with an ampoule of the deadly virus.

The chimpanzee, a 130-pound male named Doobie, had been anesthetized and was strapped to an examination table, ready to receive the injection. Everyone waited with bated breath as Professor Girard handed me the syringe that contained about one thousand times the infectious dose (the amount of live virus calculated to cause infection in a human being). The technician, Terry Kowalski, had tied a tourniquet around Doobie's upper arm, and the vein stood out prominently beneath his shaven skin. As I approached Doobie to make the injection into the vein, I suddenly thought,

What if I screw up? What if I push the needle too far, so that it goes in one side of the vein and out the other, and the virus leaks into the tissues instead of discharging into the bloodstream? The study would be ruined. What if I accidentally push the needle all the way through the skin on one side and out the other, and I end up squirting the concentrated virus all over the place, or worse still, spraying it over Terry? I suddenly became a nervous wreck, and my hand began to shake uncontrollably. The more I tried to steady myself, the worse my shaking became. I could feel Professor Girard's eyes boring into the back of my neck. He was no doubt wondering, in his Gallic arrogance, if I had been into the whiskey bottle the night before.

AIDS research came quite late to LEMSIP—not until 1988. Even in that not-too-distant past, there was a great deal of hysteria about AIDS: surgeons were refusing to operate on HIV-infected people, nurses were walking off wards that contained AIDS patients, and many of the bars in New York City served beer in the bottle because patrons feared contracting the disease from improperly washed glasses. Concerns among the staff at LEMSIP—over the prospect of infecting chimpanzees with the AIDS virus and having the animals spit on them or, worse still, bite or scratch them—led to three senior workers quitting; one, a vet, because he was outright scared of the disease, and two technical supervisors because they were afraid that their friends and relatives would shun or ostracize them.

But the problems of working with AIDS in the laboratory were still far in the future when I made my first visit to LEMSIP in the mid-1970s, while I was still working at the primate center in Madison. I had heard a great deal about LEMSIP over the years, and about its founding director, Dr. Jan Moor-Jankowski, whom I had met once previously when he had visited Downe, in England. I had wanted to visit Moor-Jankowski's lab since I had first arrived in the United States, but my excursions to other laboratories on the East Coast—to Boston, Washington, Atlanta, and the like—never seemed to take me anywhere near the New York area. It required special effort on my part to make a short visit.

Moor-Jankowski, or M-J as he is known by supporters and critics alike, is a strikingly tall, somewhat intimidating, aristocratic-looking Pole who dresses immaculately, no matter the occasion. He came to the United State in the mid-1950s after first getting his medical degree in Switzerland and then working at Cambridge University in England. He delights in recounting stories at the expense of his many adversaries. When he worked at the National Institutes of Health in Bestheda, Maryland, soon after arriving in the United States, a foreign visitor to the campus asked him in passing how many people worked at NIH. "About half," he replied, without further comment.

Dr. Moor-Jankowski's mission, when he started LEMSIP, was to provide investigators, particularly physicians and surgeons who had bold, new ideas but little or no financial backing, with the expertise they lacked to carry out their research on primates. This laboratory-bench to hospital-bedside approach was quite a revolutionary concept at the time, and I'm not sure it has ever been duplicated anywhere else to quite the same extent.

Dr. Moor-Jankowski assigned Bill, one of the senior technicians, to be my guide for the day and take me on a tour of the lab to introduce me to the large variety of primates it housed and explain some of the research that was going on. In many ways, LEMSIP reminded me of the laboratory in Downe. The animal rooms were clean and bright, and the monkeys, even though caged singly, seemed alert and happy, and certainly very responsive to me, a stranger. Unlike the other laboratories in which I had worked or visited in America, where the monkeys were packed sixty or more, in double tiers of solid-sided, wall-mounted cages on either side of the rooms, LEMSIP's monkeys were in single tiers of open-barred cages, with no more than thirty or so to a room, giving more of an impression of openness. Just like the monkeys at Downe, the LEMSIP primates shook their cages and goaded me with open-mouthed threats, a clear indication that they thought I, the stranger, should bugger off and get the hell out of their territory.

Later, with the tour of the monkey colony over, Bill took me to see the chimpanzees. This would be my first encounter at close quarters with chimps. Like so many people unfamiliar with the species, I had imagined all chimpanzees were like the ones you see on television commercials—squat little things that chatter and grimace in a friendly, comical fashion. These adults were big hairy bruisers. Their hands were massive, half as long again as a large man's, their fingers twice as thick. Quite a number of them seemed to take great delight in spitting on me, not dainty little spits, but half-gallon deluges they sucked rapidly from the water containers mounted on their cages. No matter how I tried to avoid them, I found it impossible to sidestep the sprays. In no time at all, the disposable protective clothing Bill had given me was saturated, and my paper surgeon's hat dripped down my brow. In time, I would find that chimps, more than any other species I had come to know, have a profound sense of humor, and they delight in making fools of strangers.

Bill led me out of the last chimp room and up a short flight of wooden steps to a verandah outside one of the buildings. It felt so good to be out in the open air again, to feel the freshness of the breeze against my face, no longer the helpless target of the chimps' spitting. As we reached the top step, a door opened in front of us, taking me quite by surprise. A young technician stepped out, leading a two-and-a-half-year-old infant chimpanzee by the hand. He walked in the stooped, Neanderthal-like way that chimps do, and stood no more than a foot and a half tall. Upon seeing me, this little chap reached up, his fist tightly clenched, his face and lips contorted into an insolent grimace, and punched me, with all his might, in the back of my knee. It had happened again! Just as the Java monkeys at Downe had captured my imagination with their brazen inquisitiveness, I was lured, hook, line and sinker. I sensed a level of self I had not detected before in an animal. This little squirt's audacity and confidence went far beyond a pronouncement of territoriality, or a duty to protect, as a dog might behave towards a stranger: It was a resounding affirmation of intellectual superiority. Working with chimps, I imagined, must be the ultimate experience.

It took me a year and a half to convince Dr. Moor-Jankowski that he should consider offering me a job as gynecologist, obstetrician, and pediatrician for the chimpanzees at LEMSIP. Without realizing it, I was setting myself up for years of heartache and doubt. So often since, I have rued the day that little chimp punched me!

LEMSIP was different from any other laboratory I had worked in. The research was much more intense, and I found working with surgeons and physicians quite demanding and stressful; their egos were sometimes difficult to satisfy, and many of them showed little compassion towards the animals. But even I had still not come to see anything terribly wrong about keeping the animals in cages for the rest of their lives—provided, of course, they were treated with the utmost humane care.

My own work dealt with the breeding of the different species—the rhesus monkeys, baboons, and chimps, the study of their fetal development, the obstetrical care of the mothers during their pregnancies, and the rearing of the young. I pursued my interests in ovulation and pregnancy, and even embarked on artificial insemination in the chimps. I became involved in studies of fetal distress in baboons and how to manage it—very much an example of Moor-Jankowski's philosophy of lab-bench to hospital-bedside research—and initiated a very large study, also in baboons, to investigate the role of certain types of intrauterine contraceptive devices in the development of pelvic inflammatory disease in women, a condition of grave concern to the medical community at the time. There was more than enough to keep me intellectually stimulated.

Little by little, however, as the years passed, I began to be drawn into what one might call more "hard-core" research: the study of viral hepatitis in chimpanzees, particularly hepatitis B and, what was then known as hepatitis non-A, non-B. By and large, the research procedures themselves weren't hard on the animals—although they go through the same tissue and blood changes as an infected human, the chimps don't develop symptoms of the disease—and none of the studies required their being killed. The research primarily involved taking blood samples from the animals once every two weeks to monitor their virus load, measure their antibody response, and test their serum chemistries. Less frequently, a punch biopsy (a procedure that requires a large needle to be inserted through the chest wall to obtain a tiny core of fresh liver tissue) would be needed for electronmicroscopy. In human patients, this technique is performed under local anesthesia in a doctor's office, the patient fully aware of what's going on. In chimps, the procedure is carried out under a more general anesthesia, the animals totally unaware of what's happening to them. Even more occasionally, and only for certain types of hepatitis studies that required infected liver cells for tissue cultures, the chimps would have an open liver biopsy, a procedure that is undoubtedly more taxing on them in a physical sense. The biopsy required the animals' abdomens to be opened, under general gas anesthesia, to obtain a small, wedge-shaped piece of liver.

The studies lasted anywhere from six months, for hepatitis-B vaccine testing, up to two or even six years or longer for certain other types of investigation. Two of our chimps, Teapot and Phoebe, went through one study that lasted fourteen years. The chimps are robust, and they seem to tolerate the physical aspects of the research with little or no external evidence of ill-effect. But as I was to discover, the *psychological* effects were sometimes shattering.

Animals on the different types of virus studies had to be caged singly and couldn't be mixed or even housed in the same room with one another, for fear of cross-infection and contamination, which would ruin the experiment. In addition, all infected animals had to be separated from non-exposed animals. When it came to choosing young animals for study—the first always a vaccine test, the shortest and most innocuous type of study—the criteria for selection were that the animals had to be not less than two years of age or ten kilograms in body weight. But at this age, a chimpanzee is still a mere baby. In the wild, chimps aren't weaned from their mothers until four or five years of age, and they depend on a continued close association with their mothers, as well as older brothers, sisters, and other relatives, until they are around eight years old.

In the early days of testing hepatitis-B vaccines, every effort was made at LEMSIP to assure that young chimps from the same social group went into studies together, even though they had to be caged singly. In this way, each infant was at least surrounded by friends, and at the end of study they could all go back to their original social groups. With the discovery of the non-A, non-B form of hepatitis, it became increasingly difficult to meet these social provisions. Some infants and young juveniles would still be on hepatitis-B vaccine trials, while others had progressed to studies involving live hepatitis-B virus, and yet others had already been assigned to non-A, non-B studies. Some animals, like Blair and Todd, who had been together from the day they were born, became separated, not even seeing each other for several years. These separations proved particularly hard on males. Tough-guy show-offs until the crunch came, they were often transformed into whimpering masses of loneliness. Even dedicated technicians like Roxanne and Darlene, who had nursed them as babies through the crisis of being weaned from their mothers, couldn't safely enfold them in their arms and rock away their fears with kisses, because of the then-unknown risk of contracting hepatitis themselves.

But even more disturbing than the negative psychological effects on the animals was the question of what to do with them once their usefulness as research subjects was past. In the early 1980s, there had been a great deal of talk about the growing number of these "surplus" chimpanzees that medical research laboratories were beginning to accumulate. "Surplus" referred to those animals who, as far as anyone could tell, weren't suitable for any further biomedical research. Involved mainly in what was considered one-use-only research, they became a financial burden on research centers, unable to earn their continued upkeep. They were literally eating laboratories out of house and home. These chimps had been exposed to hepatitis B and had either developed protective antibodies against the virus (making them unsuitable for further research in that area) or, as with between 5 and 10 percent of them, had become lifelong carriers of the virus, posing a health threat not only to the other, nonprotected animals, but also to the human beings who, without the protection of as yet to be developed vaccines, had to manage the animals' daily care. Then along came another virus, suspected of causing liver disease. It was then termed the non-A, non-B form of hepatitis and is now known as hepatitis C. This virus, just like the hepatitis-B virus before it, was suspected of being a major contaminant of the national transfusion blood pool, and once again the chimpanzee proved to be the only model for its study. (The list of other viruses that cause hepatitis in human beings, many of which may taint the national blood supply, continues to grow and has now reached beyond non-A, to non-F.)

In the push to develop vaccines against these new forms of hepatitis, many colony managers began referring to chimpanzees as either "clean"—meaning they had not yet been exposed to these viruses in the laboratory—or branded them as "dirty" if they had, almost as though the animals themselves were responsible for their own contamination. Some scientists went so far as to refer to exposed animals as "old dogs"—not in the kindest connotation that one might attach to such a phrase, but more to give the sense of something used up and no longer whole. These scientists rarely, if ever, went to see the animals, never saw their individuality, never understood that, despite the animals' contamination, their personalities, their "chimpanzeeness" remained undefiled. I began to see this distancing from the animals and the demeaning terminology applied to them as a way for scientists to ease their consciences, by viewing the animals as much less than human.

Yet the question remained: What was to be done with these "surplus" chimpanzees? They couldn't be returned to Africa for a whole host of reasons, not the least of which being that they wouldn't be able to survive in the wild. Zoos wouldn't accept them because they, themselves, were often stuck with animals that no longer attracted public attention. Just about everyone enjoys seeing baby chimps romping around, but most people are turned off by hairy old adults who throw feces or spit water at visitors. There was only one organization in the United States at the time, the Primate Foundation of Arizona, that might have been able to provide at least some of these animals a permanent home and a life of retirement. The foundation, located in the Arizona desert outside Tempe, was started by Paul Fritz, a German immigrant, and his wife, Jo, when they took in two chimps, castoffs from zoos, whose only other prospects were sale to research or "euthanasia." Over the years, the number of animals adopted by the Fritzes grew, and they began to face severe problems of their own trying to raise the money to care for the waifs. The pharmaceutical industry, the main users of research chimpanzees, particularly in the development of hepatitis-B vaccines and the growing work in hepatitis C, were loath to provide funds for the animals' lifetime care and retirement. The industry didn't wish to go public with the fact that they were using chimpanzees for research.

Then, in 1980, a national committee was formed, under the auspices of the National Institutes of Health, to discuss the impending crisis regarding the future breeding of chimpanzees in the United States. The breeding population circa 1980—consisting mostly of animals that had been part of the early aerospace research program back in the 1960s, or that had been imported as youngsters from the wild in the early 1970s to meet the demands for hepatitis-B research—were aging and, it was predicted, would probably cease breeding by the early 1990s. It was generally considered that the majority of the young born in captivity in the intervening period, especially the males, would prove incapable of breeding due to behavioral inadequacies. This NIH committee, of which I was a member, met twice a year to also discuss the problem of the "surplus" animals. The issue came to a head in an autumn meeting. The committee, which comprised about thirty people from various scientific disciplines—veterinarians, behavioral scientists, virologists, and the like—was divided into two groups, one to discuss the breeding issue, the other to tackle the problem of the "surplus" chimps. Midway through the afternoon, the two groups were to switch topics. I was in the first group. After our first session, my group swapped rooms to meet with the group leader in charge of the "surplus" issue. As we sat to take our places for the discussion, I noticed on the blackboard a list of options for dealing with the "surplus" animals. It included transfer of animals to zoos, creation of retirement facilities, return to Africa, and so on, but number one on the list was euthanasia. The section leader turned his head to his new audience as he started to erase the writing on the board and said that he wanted to make clear to everyone that there had been no attempt by the last group to "prioritize" the list; euthanasia just happened to end up number one.

My group discussed and debated, philosophized and pontificated, cajoled and argued, sometimes heatedly, for the next two hours on the various options, and guess which came out number one on the list again? Euthanasia!

At the next meeting, held the following spring, the chairperson of the committee, a senior member of one of the branches of NIH, announced at the outset that we had all agreed at the meeting held the previous autumn that euthanasia was an acceptable alternative to dealing with the problem of the "surplus" chimpanzees. I was astounded. No vote had been taken at the previous meeting to determine the members' final verdict. I, for one, hadn't come to such a decision, and I made my objections known to the rest of the committee. Only one person came to my defense—to the chimps' defense, that is—a young German veterinarian, Jorge Eichberg, who worked with chimpanzees at the Southwest Foundation for Research and Education, in San Antonio. Stemming from discussions he and I had had over breakfast that morning, he would, a few years later, develop the concept of providing chimps with an endowment policy, to pay for their upkeep after they are retired from research, a policy that has since been adopted by a number of research facilities.

On the flight home to New York, I scribbled a letter to the chairman of the committee and another NIH representative who had been present at the meeting, imploring them not to sanction euthanasia as a solution. Once NIH, the leader of medical research in the United States, gave its blessing to the act, it would only open the floodgates to directors of research laboratories faced with the overpopulation of seemingly useless chimpanzees. Little did I realize with these words how close to home this prospect would hit. I had no way of knowing exactly how many chimpanzees were at risk of being killed—I guessed the figure ranged between 50 and 150—but LEMSIP, which housed about 10 to 15 percent of the national chimpanzee population at the time, had already drawn up its own list of 14 candidates, animals that had been infected with hepatitis C and presumed, therefore, to be lifelong carriers of the virus. I went on to give five scientific reasons why it would be unwise to adopt such a policy, listing, as the most important, the need to reserve chimpanzees if a new disease, for which only this species could be used as a model, should come along. (My remark would prove to be strangely prophetic: within two years AIDS came hammering on the door with a vengeance.) In such an event, I pointed out, if the "surplus" chimpanzees were killed, we would find ourselves having to dip into the breeding stock of the various laboratories (which were now being looked upon as a national resource), because these would be the only animals left available for use. I ended my short, hurriedly written letter on a philosophical note: We could surely do better than kill, out of hand, man's closest living relative. I probably should have confined my remarks to utilitarian issues and not mixed morality and ethics into it. Perhaps as a result, I got no reply to my letter, even though the chairman promised me over the telephone that he would respond.

I became alarmed and despondent about the situation, sure that euthanasia would become an accepted policy. Again, as with the little mice I had worked with in Wales when I was studying for my master's, it wasn't the killing so much that disturbed me—that could be done humanely, without suffering to the animals—it was the permanency of the act, the snuffing out of something exquisitely beautiful and unique, in this case the lives of animals that are our closest living relatives. Killing as an integral part of science was something I'd had to painfully accept, as often is the case in research involving rhesus monkeys. But there was no scientific reason to contemplate killing the chimps. Even the term "euthanasia" was a misnomer; the animals weren't being killed to end their suffering, for the simple reason that they didn't suffer disease from the viruses they had been infected with. And although no one would want to come out and say it in such crude terms, another reason to dispose of the infected animals was to create empty cages that could be filled again with fresh, new animals that would be much easier to work with and care for. This I found totally unacceptable.

If we could do nothing to protect the fewer than fifteen hundred chimpanzees in research—our next of kin, closer to us than they are to the gorilla and orangutan, separated from us by only four to six million years of evolution, sharing 98 percent or more of our genetic code—what chance would there be of securing some compassion and hope for other species like the rhesus monkeys, the baboons, the tiny marmosets and tamarins, not to mention the dogs and cats and all the rest of the creatures we use in research?

Yet, less than six months later, I would come face to face with the issue myself. Two of the fourteen chimps on LEMSIP's "surplus" list were assigned to an acute, terminal study. The experiment was not performed at LEMSIP, but at another laboratory where the animals were put under deep anesthesia and then killed. For all my high-minded pronouncements and imploring, I let it happen. I did not argue strongly enough to stop it, did not lie down in front of the truck to prevent the animals from leaving the lab. I was a coward, and the shame of my cowardice in that situation haunts me to this day.

Ironically, it was the AIDS epidemic and the chimps' genetic makeup that saved many of these animals from being killed. It was not by the kindness of some administration, but through the necessity of research: once again, the chimpanzee proved to be the only animal model for the study of a major human disease.

In the past fifteen years, many scientists have come a long way in their thinking about the two thousand chimpanzees now used in research. A 1997 report of the National Research Council, a committee of scientists from a broad range of disciplines, concluded: "The phylogenetic status and psychological complexity of chimpanzees indicate that they should be accorded a special status with regard to ethuanasia that might not apply to other research animals, for example, rats, dogs, or some other nonhuman primates. Simply put, killing a chimpanzee currently requires more ethical and scientific justification than killing a dog, and it should remain so." This statement does represent a change in thinking, but it's still certain to send shivers down the spine of any true animal lover.

Because of the nature of my job at LEMSIP, I began to see the animals not as members of a particular species, but as individuals, animals I had grown to know. I had brought many of them into the world, struggling alongside Mike and Dave, my two breeding technicians, to deliver their babies, or nursed them through long nights of care with Terry, Roxanne, and Darlene, when the animals suffered pneumonia, heart failure, or other debilitating diseases. I had also become the one who selected the chimps for research. I began to see this process of selection as giving the kiss of death, knowing that some of the animals risked being thrown away on the biological scrap heap once their usefulness in the study came to an end.

I voluntarily took on the role of selecting the chimps and monkeys for research, but I would have gladly disavowed myself of the responsibility. My colleagues wanted me to prepare lists of animals, showing their age, weight, sex, infectious status, and other types of biographical data, so that they could simply choose for themselves which animals would be suitable for the next study. But they didn't know the animals as individuals like I did. Above all, I wanted to make sure, once I knew what the biological criteria were, that only those animals I thought psychologically able to withstand the study were actually assigned to it.

This selection process became the most unpleasant part of my job, even though, for most studies, the mechanics of the process were relatively straightforward. An animal is almost never put on a new project fewer than six months after completion of the previous one, and it might be several years before they're called upon to go back into active service. Yet the chimps and the other primates are not inanimate objects. Their moods change over time and their needs vary. I found this to be particularly true for two- to five-year-old infant chimpanzees. But even the juveniles and subadults of seven to twelve years old or so, especially the males—going through their equivalent of the "terrible teens"—are surprisingly fragile. For all their piss and vinegar, their show-off rattling of the cage bars and their brazen spitting, they can be psychologically demolished in a moment. And then there are the old-timers, the chimps in their late thirties or older, about whom you find yourself asking, "When is enough enough?"

I would try to stall in selecting animals by saying that I had forgotten; or I had momentarily lost the list and would produce it without further delay; or I wasn't happy about the result of the last blood test, and I ought to do another one—anything to give me a chance to consider more fully which animals I should finally select.

Possibly the most difficult selection I ever had to make involved Calvin. Calvin was a gracious old gentleman of a chimpanzee: I can think of no better way to describe him. His back was bowed, the bony protuberances of his spine stuck out like a mountain range down to his pelvis. He walked stiffly, because of arthritis, and the tips of his light brown hair were bronzed, all signs of advanced age. Yet despite his age, he was an unusually happy chimp, expressing his pleasure with anyone who stopped to talk to him by performing little pirouettes and laughing in the grunting, huffing sort of way that chimps do. He insisted that you share food with him, not occasionally, like many of the chimps, but always.

Calvin had seen his share of research over the years, but he had been retired from the more demanding studies as he got older. Even so, he would still be used occasionally for short-term studies. His most recent stint—referred to as a pharmacokinetic study—involved his being restrained on a table, under anesthesia, for eight hours. While he was restrained he was injected with the drug being studied, and then blood samples were taken at frequent intervals to determine the rate of disappearance from the bloodstream, and the rate of clearance into the urine, of the drug. As studies go, this would be looked upon as fairly light, innocuous duty.

Even though Calvin remained lightly restrained by arm and leg straps throughout the day of the study, he was cushioned on a thick, foam-rubber mattress. But it wasn't until several days after the study that we realized he was beginning to develop skin sores over all the high points of his spine. Once the first signs of bedsores appear, it's already too late to stop them; the best that one can do is try to prevent them from becoming more extensive. Most of us are unaware that bedsores (which frequently affect elderly, bedridden people) can actually become life-threatening. Calvin's sores took months to heal, in spite of his patiently coming up to the front of the cage twice a day to allow us to wash his wounds and apply fresh antibiotic ointment, a degree of cooperation that few other chimps would have shown.

Then we received a request to perform a new AIDS-related study to test the protective effects of concentrated human antibodies to a virus closely related to HIV. This time, however, the selection criteria were unusually strict. Among other attributes, the animal would have to have a particular blood group factor—one that's rare in chimpanzees. Examination of Dr. Socha's blood-type records of all the chimps at LEMSIP indicated that there were only 2 animals out of the nearly 250 in the colony that carried this factor: one, a one-and-a-half-year-old infant; the other, Calvin. The infant was out of the question. Not only would we not want to subject a baby to this kind of study, but he wouldn't have enough blood for us to take the amount the study required. Yet the thought of using poor old Calvin caused my stomach to churn.

"We can't use Calvin," I said to my colleagues. "He's too old."

"For goodness sake, it's only going to be a one-day study," responded my colleague who would be in charge of the study. "I don't think that'll be too hard on him."

"Yeah, but he gets bedsores so easily," I retorted, "no matter what precautions we take. Having him on the table that long is going to be really hard on him."

"There must surely be ways you can bolster him up and give him plenty of soft materials to lie on," my colleague rebutted.

"But he shares his food with me," I returned lamely, running out of any other immediate ideas.

"Well, why don't we use him and lose him?" my colleague replied in exasperation, a comment that would long be remembered by the technicians who were present in the room. "What good is he to us if we can't use him in research?"

Certainly no one—myself included—expected a problem injecting the material into Calvin. It had been extensively tested in rabbits and monkeys, and there was no evidence of deleterious side effects to any of their organ systems. It was only the use of Calvin in one more study that would require day-long anesthesia that concerned me.

Over the next several weeks, I wrestled with the idea of using Calvin in this experiment, finally concluding that I was being far too remonstrative: we had no alternative. The study was scientifically justified, the possible benefits to man undeniable. After all, that was what our mission was all about: the development of treatments for ill people through the use of animals. If we really put our minds to it, we certainly would be able to make the conditions more comfortable for Calvin, and we probably could keep him from developing pressure sores. My earlier objections were unreasonable, I finally concluded.

The day of the study, with everything ready to go, Calvin was anesthetized and strapped to the table on an extra well padded foam-filled mattress, ready to receive the antibodies by slow intravenous drip. The port of the drip line was turned on; instantly Calvin's body leapt into the air, as though he had been jarred with a jolt of electricity.

Calvin was dead, not even a fading flutter of a heartbeat registering on his ECG trace. He had suffered a massive heart attack.

To say that everyone was shocked by this event would be an understatement. Yet the shock was due less to the loss of Calvin as a creature than to the loss of a research subject; it was a reaction to a research project gone hopelessly awry, ending a promising investigation into a new approach in the fight against AIDS. No one mourned the loss of Calvin as a unique and very special soul. No one would be saddened by his departure, except, of course, for the technicians and myself, who had known him so well.

I relate this story about Calvin not as a way of criticizing my colleagues, but to emphasize the dilemma I had got myself into. I was beginning to find it increasingly difficult to use the animals without considering the costs to them, as individuals, in spite of the benefits to man. Like my colleagues, I accepted the need to use animals for the benefit of mankind, but I was beginning to say "Yes, but . . ." too often. My colleagues were committed to their goals. I was certain that none of them would countenance cruelty, or the frivolous use of an animal. And, like any military commander committing his troops into action, they would have to calculate the casualties and determine whether they were morally justifiable. I worried that I was becoming a moral fence sitter, ready to hop off on either side whenever the occasion suited me.

Chapter Eleven

■ ■ ■

LIKE CALVIN, THE LITTLE Jamaican puppy was special in her own way, if for no other reason than she existed. She might be only one of a thousand puppies in a similar plight right at this very moment in Jamaica, yet her life was just as worthy of my consideration as those of the chimps and the monkeys back at the lab. It was going to be another long night of vigil, I was sure, because I couldn't help worrying that something could still go wrong following the transfusion. She might suffer a delayed hypersensitivity reaction if her mother's blood was the wrong type, and I couldn't rule out the possibility of an undetectable hemorrhage starting up inside her during the night. I also couldn't help thinking that in spite of the transfusion, the puppy would still die if we gave her back to Miss June, because of the risk of reinfestation with fleas, or undernourishment, or infection. On and off, over the last couple of days, I had been toying with the idea of taking the puppy back with us to America, but I was wary, knowing well that doing so would be fraught with difficulties.

The stresses of the long journey alone, for such a young animal in poor health, were daunting enough. Besides, the puppy had already been spoken for by the farmer and his mentally retarded little boy. And how would Miss June respond to my request to take the puppy back with us to America? Miss June's frequent inclusion of God in her conversation made me think that she had such total faith in Him, she might well feel that He had His designs and I should do nothing to thwart them. But, before worrying about God and Miss June, I would have to get Marie-Paule's opinion on the matter.

I knew that Marie-Paule would be loath to become emotionally attached to another dog, only to go through the same heartbreak when it died. Our dog, Angus, was about twelve years old when I had to put him to sleep because of incurable cancer of the intestine. A white, rough-coated, stocky little mutt, he had cockily wandered into our lives when we lived in Wisconsin and became a very important part of the family. Even after four years, we still missed him sorely. Equally important, however, was that life had changed for all of us since those days, and we couldn't offer this puppy the same life of freedom that Angus had had. Marie-Paule was no longer a stay-at-home mother seeing to the children, house, and the million other things that mothers attend to. The teaching job she now had required her to leave home early in the morning, sometimes not returning until quite late in the evening. This would mean the puppy's being left alone for long hours; our children, now grown, wouldn't be around to act as playmates, as they had been for Angus.

Angus had a full life. Very much a free spirit, especially

in his younger days, he had people to see, places to go, and things to do each day. His day began around 8:00 A.M., as he waited with the children for the school bus, which picked them and the neighbors' children up at the end of our small road. This duty over, Angus would return to the house to wait atop the settee in the living room for four dogs who would call by each day to pick him up. There was Taffy, an Irish setter, who loved raiding people's garbage on pickup day and would take every opportunity to slip into people's houses and steal their shoes. There was also Tasha, a large malamute, and Pinup, her live-in companion, a skinny black dog who had a particular fondness for Angus. Lastly, there was an Old English sheepdog, nicknamed "Mrs. Snoop" by our children, who lived a mile away through the forest and came every day, for years, rain or shine, to be with Angus. Once assembled, the five of them would take off into the woods for an hour or so.

By mid-morning Angus returned home, only to start on his daily rounds to visit two housing developments, one to see an old man he had befriended, the other to visit a weepy-eyed female toy poodle. Then, in the early afternoon, there was the Silver Dollar, a seedy bar at the edge of a large lake about a mile and a half through the forest from our house, where Angus would go to cadge tidbits from the patrons. For a little dog with short legs, Angus had a most demanding daily routine. But he was always home in time to greet the children as they got off the school bus.

Angus had an incredibly close relationship with the children, and he obviously saw himself as one of them. He participated in all their activities; he sat in the secret world of Nathalie's walk-in closet as she played with her dolls; he shared the bed with Pádraig while he struggled through his homework; and he oversaw Christopher's construction projects. And everyone spoke to him constantly, involving him in every conversation. It was a perfect life for a dog.

The Jamaican puppy would have none of this. Not only had the two boys grown up and left home, but Nathalie would also be leaving soon to explore the world. The only dogs remaining in the area were Asa, a staid old Rhodesian ridgeback who lived on the other side of the pond, and Baxter, a cocker spaniel who lived next door. Both, I thought, were too mature to be bothered with a little puppy.

But, despite these concerns, I couldn't help feeling that we had no alternative but to take the puppy back home with us.

I finally plucked up the courage to raise the issue with Marie-Paule. "You know, Marie-Paule," I began, "I've been thinking. Maybe we should consider taking the puppy back with us to America. There's no good saying—"

"Yes, I agree," Marie-Paule said, cutting me off in mid-sentence. "I've been thinking exactly the same."

I couldn't believe it; I had imagined I would have to go through all sorts of complicated arguments as to why we should adopt the puppy, but there Marie-Paule was, agreeing with me. I should have realized, of course: She had fallen just as much in love with this little thing as I had. "The puppy will never make it if we leave her here in Jamaica," Marie-Paule added. "Besides," she said, "Nathalie will be home awhile, before she goes off to teach in Spain. You know she would love to look after the puppy. And I don't have to return to school for another three weeks or so.

"The only problem is, you're going to have to find a way of telling Miss June," Marie-Paule concluded. She had obviously put a lot of thought into this and shared my concerns about Miss June's attitude.

"I'll speak to her in the morning," I said.

As it turned out, Miss June was delighted. "I was going to broach the subject with you, myself," she said, "but I thought you might think me presumptuous."

"What about the little boy who was going to have the puppy?" I asked Miss June. "He'll be so disappointed."

"Don't you worry yourself about that: I'll make sure Seán is all right," she replied. Later, I was happy to discover that Seán did get his puppy, though from another litter.

Having decided that we would take the puppy back with us, there was one important issue to settle: the puppy needed a name. We couldn't continue referring to her as "the puppy" or "the little thing," as I tended to call her. "I've already thought that one out," I announced to Marie-Paule. "She's the only one of the puppies that looks like her mother, and her mother's name is Molly, so let's call her Molly, too." So Molly-Too became her official name, although, we almost immediately began referring to her simply as Molly. It had a nice ring to it; besides, it happened to be my mother's name as well.

With Miss June's blessing procured, I realized I needed some help making arrangements for our return to the States. Marie-Paule and I were scheduled to return to New York on Saturday, in the early afternoon, and it was already Thursday morning. How could we arrange to get the little Molly out of the country in time? Not only was she very young, but she looked as though she was suffering from just about every disease known to man. I would need a veterinary certificate stating that she was free of infectious disease. What vet in his or her right mind would sign such a statement? Where would I even find a vet to perform the examination? The veterinarian would also have to verify that Molly was old enough to meet the minimum age requirement set by international airlines, and although I didn't know what the age limit was, I was fairly sure it wouldn't cover a puppy barely four weeks old.

I also doubted the airline would permit us to take Molly directly on board with us, insisting instead that she be put in the hold. Even though the holds for live animals are pressure controlled and temperature regulated, I couldn't see Molly surviving the journey if she was left alone in a crate for several hours in the dark, with no one to hold and comfort her, no one to feed her or keep her warm.

Chapter Twelve

■ ■ ■

THERE HAD BEEN OTHER baby animals I had put on airplanes that would take them far from home. There was Spike Mulligan, the baby chimpanzee, and the two little rhesus monkey babies, Finnegan and Erin. Helping them through the struggles they faced in their young lives had a profound and lasting influence on me.

I HAD DELIVERED SPIKE by cesarean section because his elderly mother, Bella, was suffering from what I thought was pregnancy toxemia. This common yet poorly understood condition used to be a major cause of human miscarriage, premature birth, stillbirth, and even maternal death during pregnancy. The standard treatment was pregnancy termination, regardless of the infant's chances of survival, whereupon the mother's condition usually reverted to normal. Nowadays, with the vastly improved maternal and prenatal care available, both mother and baby stand a much higher chance of survival.

Bella began to develop edema, the first outward sign of toxemia, when she was only two-thirds of the way through her approximately eight-month pregnancy. I had no hope the infant would survive if I performed a cesarean at such an early stage of development, yet I didn't want to risk Bella's life by not terminating the pregnancy. It became a desperate waiting game; each time I examined Bella, I tried to weigh the risk to her life by permitting the pregnancy to continue against the risk to her infant if I performed surgery too early. Chimpanzees, being the big, strong, and usually uncooperative creatures they are, cannot be examined properly unless anesthetized, and it concerned me that in the process of trying to help Bella, I might actually be harming her with the repeated anesthesia such frequent examination required.

Week by week, Bella's condition worsened. Her ankles started to swell, and even her lower eyelids were beginning to puff as her tissues became progressively more waterlogged. Judging from the abnormal fetal heart-rate patterns I was getting on examination, placental function seemed to be deteriorating, and the fetus was beginning to show clear signs of serious distress from oxygen deprivation. I could wait no longer. Even though it was still six weeks to Bella's due date, I would have to perform the cesarean right away.

Ever since I had discovered Bella's condition, Dr. Ray Hayes (the other vet at the lab), the technicians, and myself had been on standby, ready to perform emergency surgery at a moment's notice, day or night. We had repeatedly rehearsed each person's role in the procedure. John, the senior technician in those days, would be my skilled surgical assistant. Bill, the technician in charge of anesthesia, would have to be ready to act instantly to adjust the gas mixture if Bella's blood pressure began to fall or to administer cardiac stimulants if her heart rate declined. Of equal importance would be Dr. Hayes and Michelle, the technician in charge of infant care, who would assist him. They would work as a separate team, looking after the infant once it was delivered, making sure its airways remained clear of birth fluids, its heart rate stayed within normal limits, and its body temperature remained stable.

Once the decision to go ahead had been made, we prepared for surgery. When we had Bella fully anesthetized and positioned on the table, an IV line in place, various electronic monitoring devices attached and her skin shaved, scrubbed, and draped, I made one last presurgical scan of her abdomen to listen to the fetus's heart with a Doppler probe. I was aghast: the heartbeats were erratic, plunging irregularly into slow rhythms, a strong indication that the fetus was suffering from severe hypoxia, a life-threatening lack of oxygen. I would have to be fast if I was going to get this baby out alive, yet careful enough in my haste not to nick the placenta with the scalpel or scissors as I cut into the uterus. If this happened, I would be faced with an uncontrollable hemorrhage, and the baby would almost certainly die from blood loss. From the moment I made the incision in the mother's abdominal skin until I removed the newborn from her uterus—a critical period of five to ten minutes—I would be totally out of touch with the fetus, unaware of its condition, unable to determine if its heart had already stopped or if its brain was already irreversibly damaged from lack of oxygen.

As I cut through the wall of the uterus and then the semitransparent amniotic membrane enveloping the fetus, instead of the normal explosive discharge of crystal-clear amniotic fluid, a thick grayish sludge, flecked with clumps of dark yellowish-brown meconium, oozed out. Meconium is the infant's first fecal material, which usually isn't passed until the second or third day after birth. Its discharge before birth, into the amniotic fluid, is a sure sign that the fetus has been suffering for some time from severe oxygen deprivation. "The baby may already be dead," I announced to everyone around me. "Come on, little baby," I gasped as I struggled to fish the fetus's head out through the incision in Bella's womb. The baby's head always seems bigger to me when I perform a cesarean section than the incision I'm able to make in the uterus, and I wonder how I'm ever going to get the head through without ripping the mother's delicate uterus to shreds. "Oh come on, you little bugger," I shouted with mounting frustration and alarm. "You've got to try and help me, you know." Everyone in the room held their breath; all was silent except for the constant soft *bleep* of the electrocardiograph in the background and my intermittent outbursts of swearing. Finally the baby's head was free, and I was able to slip its body out through the incision and into a warmed sterile towel.

The infant didn't look good at all. His skin was a dirty bluish-gray color, and he was covered in a slime of meconium as thick as axle grease. The pulsations of the arteries in his umbilical cord were barely visible, and his throat was clogged with thick globs of mucus. Most alarming of all, he took not a single breath, despite my gently slapping him and holding him upside down to drain the mucus. In a panic, I yanked open his mouth and roughly swabbed the back of his throat with gauze pads to clear the mucus. Then, kneeling beside the surgical table, I pulled the mask from my face, casting aside all pretense of maintaining sterile procedure, and pressed my open mouth around the baby's muzzle, giving him mouth-to-nose resuscitation. After half a dozen breaths, the baby suddenly took one great long gasp on his own, and everyone in the room started to cheer. Like a little steam engine, he quickly took to breathing with a steady rhythm, the color of his gums and skin immediately assuming a healthy pink hue. The crisis was over. I tied, then cut, the umbilical cord and handed the baby over to Dr. Hayes and Michelle, who were waiting to do their part.

As I prepared to complete the surgery, first to deliver the unhealthy-looking placenta and then to suture the uterus, the double layer of abdominal muscle, and finally the skin, I kept calling out to Dr. Hayes and Michelle, "How's the baby doing?" But most of the time I got no answer; they were too engrossed with seeing to the infant's immediate needs to pay any attention to me. It wasn't until I heard the first coarse, croaking cry—a wonderfully exhilarating moment—that I knew baby Spike was going to make it.

Bella, on the other hand, didn't handle the surgery well at all. Her condition continued to deteriorate, and she died four days later, not from pregnancy toxemia, as I had originally thought, but from incurable degenerative renal failure, her kidneys shriveled to the size of walnuts. Spike was now an orphan.

At that time, in the autumn of 1979, LEMSIP still didn't have a nursery. In addition, the lab had just gone through a major financial crisis—not the first or the last in its up-and-down existence. In spite of having been the first in the field, LEMSIP lost the renewal of a large five-year government contract to develop vaccines against hepatitis B, because a competing laboratory was able to underbid us by 50 percent. As a consequence, LEMSIP lost over half its chimpanzee colony—that portion owned by the government, including all the infants born during the previous five years —as well as a major portion of its animal-care staff, who had to be laid off.

This left Spike to be reared alone, with no other infants even close to his age remaining in the colony. I knew the chances of normal psychological and emotional development were next to nil if I left him alone in a cage at the lab. I was also concerned that we would be unable to provide the intense physical care at the lab that an orphaned infant chimp would need. In infancy, a chimp must be fed hourly, around the clock. But care by humans—feeding, along with burping and diapering—requires only about twenty minutes of time before the infant can then be placed back in his or her incubator. This degree of attention works out to a total contact time between chimp and caregiver of about eight hours a day—far less than the continuous care and contact a chimpanzee mother provides. As a chimp cared for by humans begins to grow, the situation becomes more severe; feedings occur only once every four hours, dropping daily contact time to about one and a half hours. It's easy to begin treating a newborn chimp as helpless and seemingly unresponsive. Yet that newborn, like a human baby, has essential physical and emotional needs that only continuous nurturing can satisfy.

I couldn't abandon Spike. I took him home, against all the rules at LEMSIP. Although I had no intention of making him a pet, I wasn't sure I was doing the right thing, for him or my family.

But for all these worries, Spike was a joy. He was easy to rear at home. We all took turns carrying him around, either nestling him deep inside our pullovers, or wearing him suspended on our fronts in a human infant carrier. The constant motion of our bodies, which made him grasp with his fingers and toes at our clothing (in the same way an infant chimp learns to hold on to his mother's hair), encouraged normal reflex development and strengthened his muscles. We also talked to him constantly, to remind him that, even when he was snuggled in his incubator, he wasn't alone.

Our dog Angus acted as a dedicated surrogate mother to Spike (as he would over the years for many an infant chimp, as well as two infant rhesus monkeys). He provided the perfect blend of gentle concern and tender roughness they needed when they were young and defenseless. Angus would lie patiently by the incubator for long hours, to the point of near exhaustion, ready to alert us to Spike's first stirrings, licking him from head to toe, and turning him over with his nose for frequent baths, allowing Spike to cling to his chest or back so he could take him on long walks outside. By the time Spike reached five or six months old, a more boisterous relationship developed between him and Angus. Our kitchen, which extends the full length of the house, became the playground for their rough-and-tumble antics, as Angus dragged Spike by the shoulder of his little sweatshirt from one end of the kitchen to the other, only to have Spike break away, attack Angus, and repeat the game all over again. Life at home became bedlam. Everything was gnashing teeth and flailing arms, yet they never once hurt each other.

Even so, I knew the degree of security, love, and stimulation that the family and Angus could provide Spike wouldn't be enough to assure his growing up a normal chimp. It was then that Paul and Jo Fritz of the Primate Foundation of Arizona came to the rescue. They offered to take Spike, when he was eleven months old, to be reared for a year or so with two of their own young chimps, Aki and Tulik, who happened to be the same age as he. As a temporary solution to Spike's social needs, it was an ideal arrangement.

I took Spike, dressed in a warm zippered toddler's suit, to La Guardia Airport for the long flight to Phoenix, assured by the airline personnel over the telephone that I would be permitted to take him directly on board with me. I had explained that, being a mere infant, he would require the same constant care as a human baby. When I arrived at the airport I discovered, to my consternation, that despite my protestations, I couldn't take him in the cabin with me and he would have to go in the hold in a pet crate.

I can only guess how terrifying this experience must have been for Spike, all alone in his pet carrier. No amount of consoling him, before he disappeared on the luggage conveyer belt, could have put his mind at rest. Did he think I had just abandoned him?

We missed Spike during his year and a half away in Arizona, a seemingly unnatural state of tranquility returning to our home. I managed to visit him on a couple of occasions. He, Aki, and Tulik had become close friends, and Spike was getting the social experience he so much needed. But I knew that by the time he returned to New York he would be too old, and too big a handful to manage, to take him home again to live with us as part of the family. It would be time for Spike to take his place in the research lab, along with the other chimps.

LESS THAN A YEAR after Spike left to go to Arizona, I was to find myself having to care for another infant primate in dire need, this time a rhesus monkey named Finnegan.

Like Spike, Finnegan had started out life at LEMSIP on the wrong foot. His mother, a very high-strung and nervous young female, had looked after him quite admirably for the first five days after he was born, but for reasons that I was never able to fathom, she turned on him when he was six days old and almost beat him to death. I was again faced with a major problem: what to do with a suddenly orphaned six-day-old rhesus monkey.

LEMSIP's financial problems had continued to worsen, and by the time of Finnegan's birth, we suffered a second, larger layoff, which decimated the staff of caregivers. The laboratory came dangerously near to closing due to the lack of operating funds. It was only through the skills and determination of Dr. Moor-Jankowski that LEMSIP survived, and slowly began to get back on its feet.

Despite the trauma we had been through, there was one silver lining to the dark clouds that overhung the lab. As LEMSIP struggled to make its comeback, I was able to hire replacements for the caregivers who had been laid off during the preceding year and a half. By and large, they were a new breed, people who tended not to say "But that's not how we used to do it," when I tried to introduce new ideas about the way we should treat the animals. I was trying to develop an attitude amongst everyone that being humane to the animals was not enough, that we had to go several steps further. We had to develop a deep sense of empathy and compassion towards the animals, and try to see through their eyes the unnatural world around them, remaining sensitive to their fears. These caregivers, mostly in their twenties and early thirties, were quite familiar with the term *animal rights*. They knew that because of the pressures from animal rights advocates, the public was becoming increasingly sensitive to the way lab animals were treated. Consequently, these caregivers saw nothing particularly unusual or remarkable about the changes I was attempting to introduce. Even so, it would take me several years to realize my goals, but without the support and understanding of recently hired young technicians, like Roxanne and Darlene, who helped me build the nursery, and Mike and Dave, who assisted me in the breeding of the chimps, I could never have made it. For the moment, however, I was faced once again with having to raise an infant all alone, this time a rhesus monkey, a task much more daunting than for a newborn chimpanzee like Spike.

During the four years I had worked at the Primate Center in Madison, I had seen all too clearly the devastating consequences to infant rhesus monkeys when they're reared in isolation in the sterile, unstimulating surroundings of the laboratory. They often develop bizarre behaviors, such as rocking uncontrollably, pulling frantically at their hair and faces, and sometimes even mutilating themselves. No amount of rehabilitation can cure these psychological deficits once they have developed. I didn't want to run the risk of this happening to Finnegan. The only hope I had of avoiding it was to find another adult female in the colony to adopt him. I had often seen Steve Eisele at the Madison lab find a female who would adopt a rejected infant, but Steve had a distinct advantage. At any one time, he had from 150 to 200 females to choose from. I had only a dozen suitable females from which to try and find a willing foster mother.

I had a visual blind constructed at the entrance to the monkey room so that I could observe the animals' behavior without their being aware of my presence. Every day Monday through Friday for the next two weeks, I set about trying to find another female in the colony who might adopt Finnegan. It was heartrending to watch him day after day. As I would gently place him on the floor of a cage, he would begin to crawl towards the female, moving like a landlocked crab, cooing pathetically to her in obvious expectation. But every female either shied away from him or tried to slap and bite him, sending him into uncontrollable fits of screaming.

Not wanting to leave him alone at the lab overnight, I would take him home at the end of each day for my family to care for, "just until I can find a female to adopt him," I assured Marie-Paule.

By the end of the two-week trial, it was obvious that I wouldn't succeed, and I knew we were stuck with Finnegan, at least until I could come up with some alternative. I had thought it bad enough taking Spike home with me every night, but a baby rhesus monkey is, as my mother would have said, "a horse of a different color." Baby chimps, even up to the age of two or three months, tend to stay put, no matter where you place them. In a playpen or crib, they are hardly more ambulatory than a human infant of similar age. Baby rhesus monkeys, on the other hand, can run, climb, and jump just about from the moment of birth. Finnegan got into everything, and tested our patience to the limit. I would catch him threading bobby pins into the electrical outlets, or trying with his tiny fingers to operate the water dispenser on the outside of the refrigerator, or stealing cookies when he thought no one was looking—anything to get himself into mischief. You hardly dared reprimand him for some wrongdoing because, like an errant child, he would then only want to do it all the more. But taking into account that he was just being a normal, self-expressing little monkey, he was really very good.

Like all rhesus monkeys, Finnegan was very conscious of social status. He adored Marie-Paule, treating her with the deepest respect as a mother figure, and would never be rough or disobedient with her. But even though he regarded me as the ranking male, he treated me as a playmate and peer, just one small step above himself. The tall, lanky, and ungainly Pádraig, at fifteen, was someone Finnegan could accept as an equal. Even Nathalie, the next in line at thirteen years old, he treated with gentleness, although he'd ignore her if she tried to reprimand him for doing something wrong. Christopher, on the other hand, who at twelve was the youngest member of the family, was another matter. Finnegan felt duty bound, when I brought him home from the lab at night, to remind Christopher that he, not Finnegan, was at the bottom of the family pecking order. Finnegan would chase Christopher unmercifully around the house for ten minutes or so, Finnegan's hair standing up on the back of his neck, his head bobbing, hands slapping the floor, ears drawn back tightly against the side of his head, face drawn into a threatening grimace, the very picture of viciousness. All the while, he challenged Christopher with menacing "grr" calls, feigning to bite him—although never actually doing so. There was nothing Christopher or any of us could do to avoid or defuse these confrontations, but once they ended each night, Christopher and Finnegan were the best of friends.

Just as I had with Spike, I began to worry about Finnegan when he reached three months of age. Without another member of his own species to grow up with and learn what it was to be a rhesus monkey, Finnegan had no chance of becoming a psychologically healthy adult. Finnegan's needs were even more urgent than Spike's because rhesus monkeys develop much faster than chimpanzees, their formative years over so much earlier. Finnegan needed a playmate.

I decided, therefore, to try a little experiment. A female infant, Erin, had been born in the colony a month before Finnegan. If I could take her from her mother every day, I thought, for just two or three hours at a time, I could provide Finnegan with the playmate he needed without depriving Erin of her mother's love. A simple enough plan, or so you might think, but it would require my lightly sedating Erin's mother every day so that I could take Erin from her. But my plan had a serious flaw. Even sedated, Erin's mother resisted me with all the strength she could muster. No matter how gentle I tried to be, no matter how slowly I tried to peel Erin away—hand by tiny hand, foot by grasping foot—her mother held on to her with a grim tenacity. In the process, Erin became terrified of me, and the rest of the monkeys in the room would explode in uproar, trying to grab me through the bars of their cages. Once I finally had Erin, I would carry her, wrapped in a towel and nestled in my lab coat, to my office, where Finnegan was waiting in his incubator. I would try to spend as much time as I could in my office, sitting at my desk making phone calls or doing paperwork while I observed the two monkeys. Finnegan was always delighted to see Erin, but all she did for the two or three hours she was with him was sit curled up in the incubator, cooing lamentably for her mother. Erin spurned all of Finnegan's attempt to interact with her, pushing him away forcefully each time. At the end of each miserable session, I would take Erin, again wrapped in a soft towel, back to her mother. This would be the only time Erin wouldn't try to bite me or squirm out of my grasp. I'd carry her inside my lab coat back across the complex to the rhesus monkey rooms. Not a peep would come out of her as she snuggled calmly inside my coat, until I reached the outer door to the room where her mother was. Then, little trickster that she was, Erin would start to scream at the top of her lungs, as if I had been beating her. By the time I opened the inner door to the room, the rest of the monkeys would be out of control, screaming, barking, and rattling the bars of their cages, inciting one another to tear me to pieces if they got a chance. Baby stealing is obviously the most heinous crime you can commit in the eyes of rhesus monkeys.

I went through this same wretched routine for the better part of a month, hoping that Erin would eventually calm down and come to realize that I wasn't such a bad person and my taking her from her mother for a short while each day wasn't the end of the world. Of course animals, especially little babies who've been kidnapped from their mothers, don't see things in that light. Erin came to hate me with a passion, no matter how gentle and kind I tried to be with her, and she even resisted my attempts to give her milk from an infant bottle. I had to admit that my experiment was a failure. Reluctantly, I decided I had no option but to remove Erin permanently from her mother, in order to provide Finnegan the companionship he so desperately needed. The decision, which meant possibly sacrificing the mental well-being of one infant to save that of another, didn't sit completely well with me, but I could think of no alternative.

Thanksgiving was approaching, and I decided to take advantage of the four days off this would give me from the lab. I wasn't scheduled for weekend duty, so I would be able to devote my full attention to getting Erin settled in at home.

I shall always remember that first night, when I brought Erin and Finnegan home with me in the car. Within minutes of driving out the gate of the lab, Erin escaped from the woolen ski hat I had put her in, tucked inside my pullover, and she proceeded to run and jump hysterically all around the interior of the car. Every time I reached out to grab her, she bit me on the fingers or urinated on me as she scurried frantically across my lap. I should have thought to bring her home in a pet carrier. Finnegan was perplexed by her behavior, not understanding in the least what all the fuss was about.

I can't say that Marie-Paule was thrilled to see me appear at the back door with two rhesus monkeys in my arms. I quickly explained to her what had happened, then passed Erin to her, again wrapped in a towel. Marie-Paule went inside and sat at the kitchen table, Erin wedged in the crook of her neck. For twenty minutes, Marie-Paule sat cuddling Erin, speaking to her in soothing whispers, assuring her "everything is all right, my little one, everything is going to be just f-i-n-e." I passed Marie-Paule a bottle of milk I had warmed on the stove, and she began to coax Erin to suck from the nipple. To my utter amazement, it worked: where I had spent close to a month trying to establish a trusting relationship with Erin, my wife had succeeded in a few minutes.

The next morning, still half asleep, I stumbled downstairs to find that Marie-Paule had already fed all the animals —Finnegan, Erin, Angus, and Girlie (our cat)—and had eaten her own breakfast, too. Before making my breakfast, I made my way to the end of the kitchen, which leads to a half bathroom on the right. To the left is a recessed area that houses the washer and dryer. As I approached the bathroom, I discovered Erin hunched on the floor of the laundry room playing with a clothespin she had found. Without intending to, I had trapped her in this narrow, dead-end corridor. She had nowhere to go, no easy way to escape. Looking up to see me—her nemesis, the one she detested and feared—towering above her, she started to scream, her face drawing into a grimace of deep concentric furrows around her cheeks and eyes.

Seeing this mask of terror on Erin's little face reminded me of the Harlow monkeys I had known at the Primate Laboratory in Madison, where I worked as the attending veterinarian during my stint at the nearby Primate Center. It was at this lab that the late Dr. Harry Harlow had studied the effects of extreme isolation on the emotional development of rhesus monkeys. Harlow placed newborn monkeys in what he himself termed "pits of despair": solid, metal-sided cages where the inmates, kept in total darkness, could neither see nor hear other monkeys for weeks or months at a time. Every time the sliding door to the box was opened at feeding or cleaning time, the incarcerated infants would implode into self-clasping balls of screaming helplessness.

These extreme studies had come to an end long before I got to Madison, but I became very familiar with the studies' former subjects. When I would walk into a room of monkeys, I could immediately pick out these animals based on a facial quality which, although I couldn't define it exactly, was very striking. Seeing Erin's contorted face brought it all back to me. Her fleeting expression echoed the stricken masks I had seen on the monkeys in Madison.

That first weekend in our home must have been a nightmare of terror and confusion for Erin. I had suddenly plunged her into an unfamiliar environment. Marie-Paule was the only warm mother figure to whom she could turn. But, blessing in disguise that it turned out to be, Finnegan quickly became a major source of security for Erin. No longer treated with disdain, Finnegan became a brother to Erin, someone she could turn to for solace, someone to look up to and emulate. By the end of that first weekend, Erin began to see that even I wasn't as cruel and terrifying as she had once thought, and we began to develop a close bond.

Now it was two little monkeys who sat on my shoulders every morning as I stood before the mirror trying to shave. They would jump down every so often to paddle in the sudsy washbasin or have a toothy wrestling match, covering themselves and me in shaving cream. It was two little monkeys who would scramble up my legs and abdomen in the shower, their clawlike fingernails gouging into my flesh as they clambered onto my shoulders to continue their little tussles, squeaking when they got soap or shampoo in their eyes. But they were the sweetest-smelling baby rhesus monkeys you could ever wish to meet.

From day one, Erin was both infatuated with and terrified of Angus. Needless to say, she had never seen a dog before. She would sit and watch Angus from a distance, chattering to him coquettishly in almost inaudible tones, but would skelter off, screaming, at his first movement. This would change, in time, and I was fortunate enough to be present when it did. One morning, Angus was stretched out on the kitchen floor in deep sleep. Erin approached him and began to circle him, chattering in her delicate voice, smacking her lips in a grimace of fear, but getting ever closer. I knew full well this would be it, the moment Erin would finally pluck up the courage to make physical contact with Angus. Sure enough, Erin leapt into the air, landing full square on Angus's chest, and proceeded to groom the long hairs of his ears and neck. Angus looked up and gave a great yawn. From then on, Erin was bonded with him and a most gentle relationship developed—quite different from the rough-and-tumble relationship Angus and Finnegan had formed.

Angus, the big hunter who never caught anything, loved to come to the lab weekends when I was on duty. This gave him the opportunity to explore the surrounding woods and sniff out woodchucks. Erin would go on these forays with him, straddling Angus's back while she held onto his scruff for dear life. Off Angus would charge into the forest, careering through the underbrush, Erin flattened against his back. They'd return, breathless, half and hour later, Angus's pink tongue lolling out the side of his mouth, Erin still clutching him in absolute terror.

Although Erin had been named soon after she was born, the technicians had never gotten around to naming Finnegan. For the first two months of his life, we referred to him as "the baby" or, if he disappeared from sight, leaving me to wonder what sort of mischief he might be up to, I might ask, "Has anyone seen that little bugger lately?" In the early days, there had been no urgency to find him a name. After all, we didn't expect to keep him very long. When we realized he was becoming part of the family, we toyed with all sorts of names, but none seemed to fit. Then one day I was watching Nathalie play the piano with the little monkey sitting on the keyboard (one of his favorite pastimes) fiddling with a piece of thread he had picked up somewhere. Nathalie was struggling through some Irish jigs and hornpipes. She seemed to be having particular difficulty with "Finnegan's Wake"—a traditional tune that, for me, always conjures up the picture of a wild, heathen race spoiling for the next good fight. Watching Nathalie and the baby monkey, so engrossed in their tasks, it hit me. What better name for the baby than Finnegan? Imagine my surprise when the next day I discovered that the female I had selected as Finnegan's playmate and tried so desperately to socialize him with had been named Erin by Roxanne and Darlene, the supervisors of the monkey colony. Erin is the anglicized version of the ancient name for Ireland. Finnegan and Erin—two names that, together, held the sound of childhood memories of faraway Ireland.

I HAD NEVER INTENDED to treat Spike, or Finnegan and Erin, as pets. Not only do nonhuman primates make unsuitable and even dangerous pets, but such a relationship isn't in the best interests of the animals. Long associations with human beings, especially in the confines of a home, result in the animals' only becoming misfits, unable to integrate with other members of their own species and unable to be true members of human society either.

Yet first with Spike, and then even more so with Finnegan and Erin, I began to realize that the ultimate cruelty that I could inflict on animals in the research laboratory was to deprive them of a decent childhood, to prevent them from realizing their potential as wild animals, free to be themselves. Prison must be bad enough for an adult, one day of confinement blending uneventfully into the next. But imagine the situation for a child. Every day counts, and there's no way to make up for a childhood lost. I often tell the technicians that the most important day in the life of an animal is the first day; the second most important day is its second day; and so on. Incredible things happen during the birth process of primates, as can be seen on the videos that Mike, Dave, and I have taken over the years, during our many nights of attending the chimps in labor. Faster than the human eye can see, but captured by a slow-motion replay, the whole stunning process is revealed. The crouching mother reaches down between her legs with one hand and guides the baby's head. Once the head is clear, she grasps the baby's shoulders with both hands and sweeps its body through her legs and towards her breast in one graceful arc of motion. The baby's arms and legs fly outwards during this sweep, the fingers and toes outstretched, and at the same instant the mother brings the baby into contact with her breast, the baby's fingers and toes clench the hair of her chest and abdomen in utter determination. It is like witnessing an exquisite ballet movement perfected without any prior practice. Then the mother's attentions begin: hugging her baby, engulfing its whole head in her open mouth, cleaning the corners of its eyes, checking its bottom, and, yes, even stretching out its fingers and toes, presumably to check that they're all there. How important are these first maternal interactions to the future normal development of the infant? What are the consequences if the infant doesn't receive them? I can't help thinking they are profound.

I tried to give Spike, Finnegan, and Erin back some measure of the natural childhood that had been denied them. Each day around noon, when the weather permitted, I would take my adopted infants deep into the woods around the lab, out of range of the paging system's constant demands. "Dr. Mahoney, please come to Unit 3," or "Dr. Mahoney, you've got a call on 2922." There, Spike and I, and then a year or so later Finnegan, Erin and myself, would find a broad, flat rock where we could sit and share lunch.

Lunch, which never really changed over the years, consisted of a tuna fish or ham sandwich, which the animals never liked but always insisted on trying, yogurt, fed from a spoon, a banana, orange juice, and for dessert, the rich fruitcake that Marie-Paule made—a mouthful for me, one for Spike, or in the case of Finnegan and Erin, one for each of them in turn—with the order of the menu never changing. This was part of my technique for maintaining discipline, and ensuring that they would remain attentive to me.

It was in these woods where Spike saw his first snake, his high-pitched alarm hoot echoing eerily through the forest as he held me tight across the shoulders with one arm, his hair erect with fear. It was here that he touched his first hairy caterpillar, cringing at the feel of it. It was in these same woods that I saw Finnegan and Erin gradually gain the confidence they would one day need for their eventual return to life with other monkeys.

When lunch was over, they would wander far into the forest to climb trees and strip the branches of their leaves, leaving me alone for a moment to contemplate the future. What was I going to do with these little animals? I had shown them a life of freedom—not the freedom, admittedly, that Finn and Erin might have had in their native India, living as part of a wild troop of monkeys, or Spike in some tropical forest in Africa, but a type of freedom nonetheless. Could I now put them back into the lab, to be the next in line for research? I could see myself spiriting away Finnegan and Erin, pretending they never existed, erasing their names and identification numbers from the computer lists. But a chimpanzee would be another matter. Chimpanzees cost anywhere from eight thousand to twenty thousand dollars. No, I would never be able to find a way to make Spike disappear. What was so special about him, anyway? I asked myself. Why should I feel I should protect him, only to expose another young chimp in his place?

In reality, there was nothing special about any of them, nothing that set them apart from all the other animals that sat in their cages while we were out enjoying the woods. But I had brought Spike into this world, given him his first breath of life. The bond we forged was so close that I couldn't begin to describe its intensity. Then there was Finnegan—macho little guy that he was, so conscious of his male superiority—and Erin, the quintessential female, so dainty and sedate, who had hated me with passion in the beginning and now was my friend. As I would watch them play in the woods, engrossed in their own activities, oblivious to my presence for the moment, I wondered where this would all end. Somehow, I felt it was I who owed them something, not the other way around. I began to feel it was like leading a blind man by the elbow across a road. He has put his trust in you, and you can't stop halfway, and say to him, "I got you this far, now you are on your own." I had to see this commitment through to its end.

I managed to find a home for the two rhesus monkeys in the Fejuvary Zoo, in Davenport, Iowa. Unlike the chimps, which become projected candidates for future research from the moment they are born, I hoped no one would ever notice the disappearance of Finnegan's and Erin's names from the rhesus monkey list. After their initial trials and tribulations, especially severe for Finnegan, Erin became the top-ranking female of the troop, proving herself to be an excellent mother, and Finnegan rose to become the alpha male, and a very protective father. I was as proud as any parent could be to witness their accomplishments.

For Spike, however, although I tried to provide for his future outside the world of research, life would ultimately not be so kind.

Chapter Thirteen

■ ■ ■

I COULD STILL RECALL THE sad feeling of seeing those infants being placed on an airline conveyor belt as they departed for lives outside the lab—Spike in his red one-piece suit, Finnegan and Erin huddled together, Finnegan holding his little woolly red bear. All three babies had looked at me as though I'd abandoned them. It was a wretched experience. My being on the same plane with them was little consolation. If anything went wrong, I wouldn't know about it, and I wouldn't be able to get to them even if I did. That little Molly might now have to face the same experience all alone made me sick with apprehension.

I knew some of the U.S. laws concerning the transport and importation of animals. I knew, for example, that dogs don't have to go through quarantine on arrival in the United States, as they would entering the United Kingdom or, ironically, Jamaica, because of the fear these countries have of introducing rabies. Jamaica, being a small island country, lives in constant dread of this deadly disease. This isn't only because of the large number of free-roaming dogs and cats, but also because of the ubiquitous mongoose, a species that was introduced into Jamaica from India during colonial times to rid the sugarcane fields of venomous snakes. These mammals would all act as natural reservoirs for the virus, and there would be no hope of ever eradicating the disease once it was introduced. The one thing I knew for sure was that although Molly might have all sorts of medical problems, rabies was not one of them.

I sat for some time pondering the problem of transporting Molly. Then I thought of the one person who might be able to help me, or at least offer sensible suggestions—Kathi Travers, who worked with the American Society for the Prevention of Cruelty to Animals in New York City, the "Big A," as she calls it.

I had known Kathi for several years. We first met while Roger Caras, who had recently been elected president of the ASPCA, was showing me around the society's new facilities in upper Manhattan. Kathi always displays a ready smile and an overflowing effervescence, and it would be impossible not to immediately fall in love with her. Much of Kathi's time, in those days at least, was spent at Kennedy Airport, rescuing illegally imported animals, especially exotic species such as parrots and monkeys. She often took these animals home with her temporarily, to give them the immediate attention they so much needed, while she set about the long and often laborious job of trying to find them suitable permanent homes in a zoo or sanctuary somewhere in the United States. I had helped her out on a couple of occasions with monkeys that needed special veterinary care, and she had helped me a few years earlier by finding a home for two little squirrel monkeys that I wanted to spirit away from the lab after they had been used in a short research project.

My problem was figuring out how to call Kathi. I didn't have her telephone number on me, and even if I had, I couldn't use my telephone card to call her direct because the telephone company would accept only collect calls from Jamaica. Besides, I doubted the switchboard at the Big A would accept a collect call from me, anyway. The only solution I could think of was to call Nathalie collect at home and ask her to find some way of getting Kathi's number, call her, explain the situation I was in, and ask her whether she would mind calling me back in Jamaica. Nathalie always loves a little intrigue, and she agreed to give it her all.

An hour or so later, the telephone rang; it was Kathi. "Hey, long time, no hear!" she said in her typically ebullient fashion. "What's up?"

"I'm almost at my wits' end," I said, and began to outline the problems with Molly, explaining that if we left her behind in Jamaica, she would surely die. I told Kathi that I was so concerned that we wouldn't get permission from the airlines to transport Molly, she being so young and obviously weak, that I had even considered, in a moment of weakness and despair, secreting her on my body, hidden under an oversize pullover, in the hope that I could smuggle her through airport customs and immigration. "Oh, please don't do that, Jim!" she implored me. "Give me a couple of hours to see what I can do at this end."

"Listen, Kathi," I went on, "there's no way I can even get hold of a pet carrier to take Molly on the plane. We're staying in the middle of nowhere, and I'm certain I wouldn't find a shop in Black River that sells such an item. To make matters worse, Marie-Paule and I have to return in two days' time, and I don't want to wait till we get to Montego

Bay to try and buy a pet carrier, and then be unable to find one."

"No problem," Kathi replied with a mischievous laugh, "I'll get someone at Kennedy to send you a carrier; they owe me one."

"Don't forget," I emphasized, "make it the smallest crate there is: Molly is very tiny. Another thing, Kathi," I added, "there might not be enough time to have the pet crate sent all the way through to us at Treasure Beach. It would be safer to have it dropped off at the airport in Montego Bay, and we could pick it up when we get there on Saturday morning."

"Don't worry," Kathi said. "I'll see to everything."

"Well, there's one other small thing that concerns me, Kathi," I pleaded, afraid that I would exhaust even her infinite patience. "The airline might not allow us to take Molly with us in the passenger cabin, and it would be terrible if she had to be put in the hold."

"Hmm," Kathi replied thoughtfully. "Let me see if I can talk to someone in the airline and try to get a waiver of the rules."

With the memories of Spike, Finnegan, and Erin still fresh in my mind, I almost told her not to bother. Yet I had often seen people taking their toy poodles and cats with them on board, at least on Air France flights, and I remembered a woman recently telling me that on her flight back from Madagascar, the passenger sitting next to her had suddenly produced a bloody great chameleon, who proceeded to crawl his way around her ankles. Maybe Kathi could work wonders for me after all.

Late into the evening, Kathi called me back. It had taken her longer than she had anticipated to make the various contacts. What a gem she was; all this effort just for a little runt puppy. "The pet carrier is no problem," she informed me. "I'm having it sent through to the ticket counter at the airport in Montego Bay. You can pick it up when you check in on Saturday. The freight and customs charges have all been taken care of, so you shouldn't have any problems there. I spoke to someone at the airline, and they said that you can take the puppy with you on board provided you clear it with the pilot beforehand."

"Well, I'm sorry to be a pest, Kathi," I said, "but there's one more complication I've just thought of. I'm going to need a veterinary certificate, but I haven't the faintest idea how to go about finding a veterinarian who will sign one for me."

I recounted the problems I had had with transfusing Molly, and my failure to find a local vet. "Well, you're a vet, aren't you?" she countered. "Why can't you write your own health certificate?" I hadn't thought of that. I was still a member of the Royal College of Veterinary Surgeons, licensed to practice veterinary medicine and surgery in the United Kingdom and, by extension, in the Common Market countries of Europe, even after all these years of living in the United States. After all, I told myself, Jamaica is part of the British Commonwealth; perhaps the authorities would accept a certificate written by me. The trouble was, I had nothing on me to prove that I was a vet.

In the middle of all the telephone calls back and forth with Nathalie and Kathi at the ASPCA, I received word of another animal in trouble, this time from Kathy Granger, the business manager at LEMSIP. She apologized for disturbing me on vacation, but said that Janis Carter had telephoned from the Gambia and needed to speak with me urgently about one of chimps, Johnny.

Since my first trip, ten years earlier, to help Janis relocate Lucy and the other chimps on Baboon Islands, I had been back to the Gambia only once but had stayed in close telephone contact with her on how to handle the occasional medical problem that arose with the animals.

Janis had experienced problems with Johnny several months earlier, when he had been very ill with pneumonia. Prolonged treatment with antibiotics seemed to clear this up, however, and his health appeared to have returned to normal. Then, beginning a few weeks before I was due to go to Jamaica, Johnny's condition had deteriorated, and this time he was showing symptoms that indicated he might have a partial blockage of his esophagus or stomach. Even though his appetite was normal, he would start to retch every time he swallowed and then go into uncontrollable spasms of coughing. His condition seemed serious, and it sounded to me like he would require surgery. Janis had called a few weeks earlier, asking me to come to the Gambia as soon as possible to see if I could do something for him.

Of course, treating a chimpanzee who lives in the wild isn't like treating someone's pet dog or cat in a vet's office. Janis had big challenges to overcome before I could fly over to operate. First, she and her assistants would have to build a cage at the base camp on the bank of the river to house Johnny before the surgery and for the long period of convalescence thereafter. The cage would have to be large enough to house him comfortably for a prolonged period of time and secure enough to hold him—a fourteen-year-old adult male who weighed about 120 pounds and had the strength of two or three men. Acquiring the construction materials would be difficult enough; needless to say, you don't find a conveniently located hardware store or metalwork shop in the bush, and such specialized building supplies have to be shipped in from the United States or Europe, at vast expense.

Janis's next problem would be to find Johnny on the island and blow-dart him with a tranquilizing agent, a difficult undertaking at best. For all the apparent seriousness of Johnny's condition, he was still quite active and able to run fast and climb trees. If Janis failed to hit him with the first dart, Johnny would likely take off into the bush, and it would be just about impossible to find him and get close enough to try to dart him again. Then there was the danger of lowering him into the small boat to bring him over to the base camp. To top it off, everyone would have to keep a constant lookout to make sure none of the other animals was coming to Johnny's assistance, ready to attack Janis's team from behind. All of this would take time and patience —an enormous amout of time, and a great deal of patience. But that still wouldn't be the end of it. Janis would then have to build a makeshift surgery, find a table that could be used for the operation, and set up lights. Having made these arrangements, Janis would have to organize ancillary medical help, should I need it in order to carry out blood tests or other diagnostic procedures.

The arrangements were interminable, the delays unending. Marie-Paule and I had finally given up on waiting and gone on vacation, as planned long before. Now, all of a sudden, it seemed Janis might have made some headway.

Just as Nathalie had acted as the intermediary between me and Kathi Travers at the ASPCA, Kathy Granger now acted as my liaison with Janis, transmitting messages back and forth. I had already planned to take Terry Kowalski, LEMSIP's senior technician, with me, but until I knew whether Janis had finished her preparations, I couldn't set about booking a ticket for Terry and me from New York to Dakar, in Senegal, and then on to Banjul, in the Gambia. Never knowing when to expect the next call from Kathi or

Kathy, I spent most of my time waiting by the telephone. I might as well have stayed at the lab, for all the vacation I was having in Jamaica.

Despite all the complications, our last couple of days in Jamaica were enjoyable, and Marie-Paule and I were able to have a good time. Molly was gaining strength and we could even take her, snuggled in her wicker basket, along for walks on the beach. I called the airport on Friday to confirm our flight and verify that the pet crate had arrived. Everything seemed set for the journey back to the States.

About nine o'clock on Saturday morning, we said our good-byes to Miss Polly and began our journey west through Black River to Ferris Cross, then north through the steep, rugged mountains past Whithorn and Montpelier, and on to Montego Bay. Halfway up the mountains, we paused by the roadside for a quick pit stop and a welcome chance to clear our lungs of the diesel fumes belching from the line of lorries that had preceded us at an excruciatingly slow crawl up the hairpin turns. Molly squatted to urinate, her little body wobbling in the tall grass, while I aimed at the bush in front of me. I looked across the mist-shrouded valley to the rugged mountains beyond, and then down at her, tottering at my feet, and realized only at that moment that she and I were bound to each other, about to start a new life together.

We arrived at Sangster International Airport in Montego Bay around eleven, a good two and a half hours before takeoff—plenty of time to make all the necessary arrangements. We drove straight to the departures building. Before we unloaded our suitcases, Marie-Paule suggested I stay with Molly while she went to the ticket counter to retrieve the pet carrier Kathi had promised as well as make sure our tickets were in order. About ten minutes later, Marie-Paule reappeared, greatly flustered, to tell me the pet carrier wasn't at the ticket counter, as expected, but was in the freight area; we'd have to pick it up before our tickets could be processed. "We'd better get a move on," she said.

We circled the front of the airport until we located the freight area, which was off to the left, surrounded by a high cyclone fence. It was like a vast parking lot, ominously empty, with warehouses along both sides; except for a grumpy-faced security guard on duty, there wasn't a soul in sight. The very silence and emptiness made me uneasy, and I sensed things weren't going to go well. The guard unlocked the wire gates and let us in. Slowly we toured the buildings, Marie-Paule straining to read the signs on each set of metal doors, loading bays, and metal-banistered stairways. None seemed to indicate the building we needed.

I parked the car in the shade of a warehouse, and Marie-Paule took off along one side of the buildings while I explored the other, testing each door to see if it was unlocked, in the hope of finding someone inside who might be able to help us. Eventually, Marie-Paule struck oil; she found an open door that led to an untidy office where a lone woman sat at a desk sorting through paperwork.

Marie-Paule began to explain our situation. "It's Saturday," the woman responded. "No one works on Saturdays. All the freight offices, and the Customs and Excise Department, are closed. I'm sorry, but you'll have to come back on Monday."

"Monday?" I said incredulously. "But we've got to catch a flight back to the States in two hours. Nobody told me when I phoned yesterday that the offices would be closed today."

"No problem," she replied, in delightful Jamaican tones

of eternal hope. "I'll call through to the departure terminal; I'm sure I can get hold of Mr. Smith, the airport manager. He'll know what to do."

I glanced at my watch. It was 11:45, barely an hour and three-quarters till departure. There was nothing we could do for the moment except go back outside and prepare some milk for Molly, let her have a little walk and do a last-minute pee. Marie-Paule began to get really agitated. We still had to finalize our tickets, fill out the unavoidable forms allowing Molly on the flight, and then return the car to the rental agency, which was a quarter of a mile away, outside the airport complex. But unless we could retrieve the pet carrier locked somewhere in the Customs and Excise building, the whole exercise would be academic.

The minutes passed and our agitation increased. Marie-Paule returned to the office to inquire if the lady had been successful in tracking down the airport manager. "No," she said. "Departures is so busy that no one seems able to locate him."

"Why don't you go, Jim," Marie-Paule said to me with mounting anguish. "Maybe you can find him."

Once through the wire gate, I drove around the one-way system of ramps and side roads to the parking lot in front of the terminal. I rushed to the departures building, asking the policeman on duty outside where I might find the manager. "I just saw him go to the arrivals building," he replied, pointing some distance farther along the sidewalk.

I dashed down the crowded sidewalk to the arrivals building. Another policeman held me back at the glass entry doors with his raised arm. "You can't go in there, mahn," he said. "But I have to," I pleaded, "I have to find the airport manager."

"Okay, mahn," he replied with a smile. "Go see that lady

in the blue uniform," he said, pointing off into the crowd. "She'll find the manager for you. But don't go beyond the baggage examination tables," he cautioned. I thanked him, then studied the milling crowd before me. Throngs of blanched passengers, already dressed in skimpy summer clothes, waited to clear customs, while baggage attendants maneuvered heavily laden carts and customs officials checked papers. In the distance, I spotted the woman in the blue uniform and pushed my way through the vacationers.

To my relief, the woman acknowledged me with a warm smile. I began to explain my problem and told her that I was trying to locate the airport manager. "Oh, dear me," she said. "Mr. Smith left ages ago to meet you at Customs and Excise."

Now I really started to panic. Back to the parking lot I ran. I found the car, paid the ticket, and then tried to work out how I was going to get back to the freight area. It was now under an hour and a quarter to departure, and we were no further along than when we arrived. I drove up to the gate at the freight area. The security guard was sitting in his little hut listening to reggae music on a radio. He was obviously getting fed up with having to open and close the gate for me (and who could blame him), and he seemed determined to make me wait as long as he could while he searched his large keyring to find the key for the gate.

I could see Marie-Paule off in the distance, standing outside one of the warehouses beside a man I presumed to be the airport manager. "Where have you been?" Marie-Paule grumped at me. "What took you so long?" We were both beginning to get short-tempered from the stress. We hadn't had a full night's sleep for ages, what with having to get up with Molly for feedings. The heat of the sun reflected off the buildings and surrounding asphalt was almost unbearable.

Marie-Paule complained of feeling a little sick, and my clothes were soaked with sweat.

Mr. Smith led us into the coolness of the vast, hangar-like warehouse. Row upon row of wire cages extended to the high ceiling, each packed with boxes, crates, and baggage of every description. Somewhere in this mass is a pet carrier, I thought, and we'd better find it quickly or there will be no flight back to America today.

Mr. Smith methodically consulted his clipboard, to which was attached a long typewritten list of items, then read the numbers on the padlocks of each wire gate till he found the right one. There, amongst the endless piles of boxes and crates, was a huge cardboard box, two and a half or three feet long, almost as wide, and about a foot deep, with the words PET CARRIER emblazoned on it. "But this can't be the right one," I said. For some reason, I had expected to see a plastic pet carrier, not a cardboard box. And, judging by the size, I felt certain the carrier inside would be large enough to hold a greyhound, not a mite barely larger than my hand.

"No, this is it," Mr. Smith confirmed, rechecking his list. I studied the tiny puppy in Marie-Paule's arms, then looked at the cardboard box. How would we ever be able to transport Molly in such an enormous box without her getting thrown all over the place?

I lugged the box from the freight cage, carried it into the heat of the parking area, and set it on the asphalt. I yanked open the flaps and began to withdraw the contents. The carrier required one of the do-it-yourself kind of assembly jobs I'm usually not very good at. The top and bottom were wedged together, and there was a separate wire door and plastic water and food dish. There were several plastic bags containing nuts, bolts, and washers, and a booklet explaining how, in a few simple steps, the whole thing could be put together.

Marie-Paule signed the papers (the customs duty had already been paid, just as Kathi had promised), and I began assembling the carrier. It was huge, no doubt about it, and just about big enough to contain a large spaniel, if not a greyhound as I had first suspected. Marie-Paule and I stood there staring at the container (which I had actually managed to assemble without incident), trying to think of how we could safely pack it to restrict the space and prevent Molly from being thrown around during the flight. Of course, I still hoped that we'd be given permission to take Molly on board with us, but I couldn't bank on that. The carrier was too big to fit under the seat, that was for sure. We ended up opening our suitcases and retrieving towels, several pairs of trousers, pullovers, and small cardboard boxes we could fit inside the pet carrier to leave a space soft enough to cushion any blows and small enough to hold the puppy securely in place.

It was well past noon by now, not much more than an hour before departure. Marie-Paule and I repacked our suitcases and put them, along with the pet carrier, back in the car. For one last time, the guard unlocked the wire gates, and we made our way, at top speed, to the departures building.

"You stay and look after Molly," Marie-Paule said, "while I return the car to the rental office." Twenty minutes later she was back, and we joined the long queue of passengers at the ticket counter. The departure lounge was pretty full by now, as several flights were scheduled to leave about the same time.

I decided to let Marie-Paule handle everything from here. When it comes to officialdom, papers to fill out, and declarations to make, Marie-Paule is so much better than I. She seems to be able to cut through red tape and extract the most sympathetic response from even the grouchiest of bureaucrats. Even so, I had a bad feeling that things were not going to work out, that maybe there was no room on the plane for Molly, or that she would have to go in the hold, or some other unforeseen complication would arise.

I let Molly wander around the floor as we waited in line to reach the ticket counter. She was an immediate success; several little children came up to pet her and try to tickle her tummy. My only concern was that she would do a big pee, or something even worse, on the granite floor.

When we reached the counter, the agents remembered Marie-Paule from earlier in the morning and rapidly set about processing our tickets and checking our passports and green cards. "Do you have the health certificate for the puppy?" the agent inquired of Marie-Paule. She handed him the handwritten certificate which I had spent so long preparing in my neatest handwriting. I held my breath. The agent read the letter, his lips silently moving with each word, his head beginning to shake from side in disbelief. "This won't do," he said, with a sad expression on his face. "This is not a proper certificate."

"But my husband is a veterinary surgeon," Marie-Paule explained.

"How am I to know your husband is a vet? Anyway, you have to have a special certificate from the Ministry of Agriculture, signed by a Jamaica-certified veterinary surgeon," the man informed her kindly.

I knew it, of course I knew it. The agent was absolutely right; how could we have expected him to disregard formalities? "What are we going to do?" Marie-Paule cried.

"Well, I'm sorry, my lady," the agent said, with the perfect gentility that Jamaicans seem to be able to muster even in the worst crises. "You can't get on this flight with the puppy. You will have to have her examined and obtain a proper veterinary certificate."

"But it's Saturday afternoon," Marie-Paule lamented. "We don't even know how to find a vet, and I'm sure he wouldn't help us at this late hour of the day, and we certainly won't find one tomorrow, it being Sunday."

"I'm really sorry, my lady, but there is nothing we can do without the right papers," the agent said in a voice of genuine concern.

The wing of the departures hall was now empty, the last passengers long since gone to board their planes. Marie-Paule and I stood staring at each other, confused and lost. We couldn't just leave the pup, that was for sure. But we had no car, nowhere to stay for the night, and we hadn't the faintest hope, I felt, of finding a veterinarian at this late hour on a Saturday who would be prepared to give the Molly-girl a health examination and issue the necessary papers.

Seeing our consternation, one of the agents, a smartly dressed young lady, came out from behind the counter. "I could take the puppy home with *me,*" she offered. "I have a dog of my own."

"That's very kind of you," I said, touched, but at the same time concerned by the idea of leaving Molly behind. "What sort of dog do you have?" I inquired.

"She's a Doberman pinscher," the young lady answered, with a warm smile. I cringed at the thought of it: a Doberman, with the reputation of the breed for viciousness—which I knew full well is unfounded for most of them—but all the same, I had visions of the two-pound Molly being gobbled up by the seventy-pound German hound in two mouthfuls. Besides, we could not leave Molly, with all her medical problems, not to mention that we had fallen hopelessly in love with her.

"We can't do anything without a car," Marie-Paule announced. "You stay here," she said to me, regaining command of the situation, "and I'll go back to Budget and see if I can rent another car."

Half an hour later she returned, all excited. "Quick, quick!" Marie-Paule exclaimed. "The people at Budget were very kind and understanding, and they let me have the car back without charging for a new rental, and one of the young men is going to meet us outside in a moment and guide us to a vet he knows." Marie-Paule went on to explain how, on hearing her story of the puppy, the need for a veterinarian, and our missing the flight back to America, everyone had taken pity on her and wanted to do all in their power to help her out. One of the young men in the office had immediately telephoned a veterinary surgeon he knew, who had been about to leave his clinic for the day, and asked him to please wait until he brought a lady with a puppy who needed export papers to the United States. All the ticket agents seemed excited with this turn of events. "I'll wait here for you," the Continental agent said. "If you get the right papers, come back, and I'll book you all on tomorrow's flight, no problem."

I scooped Molly up, and between us, Marie-Paule and I dragged our suitcases to the exit doors of the departure lounge. The young man from the car rental office was waiting for us outside in his car, and off we set, in hot pursuit, through the mounting traffic of the Saturday-afternoon rush hour, down the main drag of Montego Bay and into the tiny, one-way side-streets on the western side of the city. Without the young man's guidance we would never have been able to find our way. I was amazed, and touched, by how many people were going out of their way to help us.

The vet's office was a small building squashed between shops and houses on a narrow, vehicle-packed side street. As we entered the front door, the heavy smell of phenol disinfectant and propyl alcohol assailed our nostrils. The veterinarian was waiting for us. I passed Molly to him, and holding her roughly in the palm of one hand, he hoisted her up in the air, and with a patronizing grin said, "She's not much of an export for Jamaica, is she?" After giving her a perfunctory once-over, not even noticing her severe eye problems, he put her down on the examination table and began to write out the health certificate, the sheet of paper headed by the all-important words MINISTRY OF AGRICULTURE.

Much relieved, Marie-Paule and I left the clinic to face the busy traffic. After a little trial and error, we found our way back to the airport. The airline agents were still there, awaiting us, just as they had promised. The man we had dealt with earlier immediately set about securing our tickets and Molly's papers for the following day's flight. Everyone was elated, genuinely thrilled that all our problems were solved.

Well, that wasn't quite true; we still had to find somewhere to stay for the night. It was three o'clock, and we had eaten nothing since very early morning, when we had departed Treasure Beach. After a quick meal in the airport restaurant, which included a dish of chicken and lentil soup for Molly, hidden on the floor under the table, we drove east along the coast road to Duncans, to see if we could find a small cabin to stay in for the night. We were successful. That one extra night of rest in Jamaica was a blessing. With the satisfaction of knowing our tickets were secured, and that Molly had a place on the plane for the Sunday flight, we could have a relaxing swim in the calm sea and go for a long stroll along the beach. Most of all, this extra time allowed Molly to recuperate. It had been an exhausting day for her, and she ate well that night and slept soundly for the first time since we had taken her from Miss June, which seemed like so long ago.

With Montego Bay only forty-five minutes' drive away, and the flight not until one-thirty in the afternoon, we were able to spend a last leisurely morning on the beach. When we entered the airport and began walking towards the ticket counter, all the agents, the same ones who we had dealt with the day before, let out a collective "Hey!"

Our one, final disappointment was that Molly wouldn't be permitted to travel in the cabin with us, because of international regulations, but would have to go in the hold in her oversized pet crate. We placed her in the carrier, and an agent carried it gently to the conveyor belt, Molly quickly disappearing through the rubber flaps on her way to the loading area. For the next four and a half hours, we would be totally out of contact with her.

We boarded the plane half an hour later, hoping that our window seat would give us a view of the pet carrier being loaded into the hold. Eventually all the passengers were seated, their carry-on luggage safely stowed, and the plane seemed ready to taxi away from the terminal. We had seen no sign of Molly's carrier. A stewardess came back to our seat, leaned over, and said to me in a low voice, "The captain would like to have a word with you, sir." What could be the problem? Marie-Paule and I both wondered, looking at each other in dismay. I stood up and made my way to the cockpit, where the captain was waiting for me. He was a great big burly man, with clear skin, rosy cheeks, silver hair, and a warm smile. "Why don't you go down and check on your puppy before she's put in the hold?" he said. "I'm scared she may not make the flight, she's so tiny."

I didn't dare tell him that I was just as concerned. As I descended the gangway, I could see Molly's crate stacked on top of a box in the luggage area in front of the plane. I made my way towards it, but was immediately stopped by an airport official who shouted at me, above the noise of the aircraft's engines, "You can't go in there, sir, it's a security area." Helpless, and not knowing what to do, I glanced up at the nose of the airplane looming above me. I glimpsed the smiling face of the pilot as he gave a big thumbs-up and a wink to the ground crew, indicating that it was okay to let me through. Molly was snuggled deep into one of the pullovers we had packed into the floor of the pet carrier, fast asleep. She was fine for the moment, but I could only pray that she would withstand the long flight. Seeing my concern, several of the young baggage attendants called out to me, "Don't worry, mahn! We'll take good care of her!"

The jet roared from the runway and burst into the cloudless blue sky. We were on our way, but for the next three and a half hours we were conscious of every bump and air pocket, every change in the drone of the engines, wondering if Molly had awakened in terror. Would she know to make her way to the little plastic dish and take a drink of milk, or would she be too terrified to move?

After we landed at Newark, we had one excruciating hour to wait before Molly's crate appeared on the oversize-baggage belt. I rushed to it and opened the wire door. Molly stood wobbling amongst the clothes. Her gums and tongue had a frightening grayish hue, and she was cold and wet from urine. Amazingly, not a drop of milk had spilled from the dish attached to the inside of the wire door, proof of the gentle care she had been given by the baggage handlers as they carried her crate to and from the plane. It was also proof, however, that Molly had had nothing to drink in over five hours. I plucked her gently from the carrier and placed her on the floor. Molly stumbled and fell head over heels. She couldn't stand up without a violent tremoring of her whole body. "Get the syringe and dextrose solution from the bag," I called to Marie-Paule. "I think she's in shock." Marie-Paule quickly filled the syringe. As I was about to let the puppy lap the sugar solution from the end, an officious-looking customs agent descended upon me. "Put that dog back in the crate!" she shouted.

"I have to give her a drink," I said. "I think she's dying."

"I don't care about that," the woman bawled back at me. "It's against the law to have an animal out of its crate in the customs area."

"To hell with the law," I shouted in retort. "What's the law to me if my dog is dying?" Marie-Paule stepped in and tried to quiet the situation. "Let's not have unpleasantness," she pleaded. It was all right; in the few seconds of the argument, Molly had eagerly lapped half the sugar solution. The crisis was over.

To my amazement, we cleared customs quickly, the officer inspecting neither our baggage nor Molly's Jamaican health certificate, not even taking a quick look at her in the pet carrier. Once outside the terminal building, we took Molly out of the crate and let her walk a little. She was very much better, her body tremoring only slightly. We gave her a saucer of the milk that Marie-Paule had made up before leaving Jamaica, rubbed her down with a towel to dry her coat, wrapped her in a towel, shoved her inside Marie-

Paule's coat, and made our way to the long-term parking lot to recover the car.

On the long drive home, with Marie-Paule continuing to give Molly sups of dextrose every now and again, her little head poking out through Marie-Paule's coat, we stopped off at a McDonald's along Route 17, in New Jersey. We ate at one of the outside tables—Molly had some more milk and water, went for a little walk, and did a great long pee. She was back to her old self. "Hurray for McDonald's!" we shouted. "Welcome to America, Molly!"

The little mite had made it.

Chapter Fourteen

■ ■ ■

AFTER I ARRIVED HOME from Jamaica, I conducted an endless string of telephone conversations with Janis Carter in the Gambia. Her chimp Johnny was becoming sicker, and she was working frantically to get ready for Terry Kowalski's and my trip to her camp to operate on him. One moment the trip was on, the next it was off; Janis would try feverishly to make last-minute arrangements for the surgery, only to discover some new complication that would again put the proceedings on hold.

Terry had been the chief research technician at LEMSIP for several years. I had known him since he was a young lad of nineteen, when I first started working at the lab in late 1977. Now I needed his assistance to perform what I suspected could turn out to be a very tricky operation on Johnny, under extremely difficult circumstances.

Janis's camp on the banks of the River Gambia comprises a collection of small, tin-roofed, open-sided, mosquito-screened huts that house herself, her stalwart Senegalese assistants Bruno, René, and Boiro, as well as, at the time of my visit, a young American Peace Corps worker named Jeff. Janis and her team had worked hard over the last couple of weeks to set up as near to a sterile surgical room as she possibly could in the African bush. To achieve this, she and her assistants had constructed a spacious cage lined with mosquito netting, intended as much as anything to keep away the flies that would swarm around once I opened up Johnny's abdomen on the makeshift surgical table. They had even installed a weak lighting system, powered through a small solar panel.

In this less than optimal setup, Terry was essential to me. I needed someone I could trust; someone who could react instantly in an emergency situation under difficult conditions; someone who could put up with my eccentricities and unpredictable mannerisms and wouldn't walk off the job in frustration, leaving me stranded in the middle of Africa. Terry and I have worked together in surgery for long enough that we can usually anticipate each other's movements and are able to just about read each other's minds. When I grunt in a certain way, Terry knows I need him to mop away the blood with a sponge or retract the tissues to give me better access to the surgical site. Because I often can't bring to mind, on the spur of the moment, the endless names of the wide range of surgical instruments, Terry also knows that if I ask for a "thing-a-me-jig," I require a retractor, or if I say, "Give me one of those whatsits," he knows to pass me a Babcock forceps. He even puts up with my occasional "Shit!" "Piss!" and "Buggers it!" as I struggle to overcome some difficulty in the surgical procedure, knowing that my outbursts are not directed at him, but at my own ineptitude. I once overheard him explaining to a visitor who was observing us perform surgery on a chimp not to be concerned by my swearing. "That means everything is going fine," he said. "It's when he stops swearing and becomes silent that you have to worry."

Terry and I had organized things as much as we could from our end before I had left on vacation. We had made itemized lists of the various drugs, supplies, and equipment we would need, making careful note of what items were in which of the six large cardboard boxes that we would carry with us. These included all the instruments we would need for major abdominal surgery, which seemed to weigh a ton. We also had several units of compatible blood, which we had drawn from four of the LEMSIP chimpanzees, in case of an uncontrollable hemorrhage during surgery. Dr. Socha had already determined the blood types and cross-matches from blood samples I had brought back from Johnny and all the other chimps in the project on my first trip to the Gambia, ten years earlier. We also included an array of antibiotics and emergency drugs and supplies.

The least part of our baggage was our own personal effects: two pairs of trousers each, a couple of long-sleeved shirts to protect ourselves as much as possible against mosquitoes and the painful bites of tsetse flies, a few changes of underwear, and lots of socks. I had warned Terry it was the rainy season in West Africa, and our feet were liable to get soaked on day one and remain soaked for the rest of the trip.

The telephone calls between Janis and me seemed endless, and by three o'clock on Wednesday afternoon, it appeared that the trip was finally off for the foreseeable future. Then, an hour later, she called back: the trip was on, the last complication overcome. Within thirty minutes, Terry and I were on our way to Kennedy Airport to catch the eight-thirty Air Afrique flight to Dakar, Senegal just three short days after my return from Jamaica.

I wasn't happy about leaving so soon. Molly was still very ill and very young, not even five weeks old, and it didn't seem fair to leave Marie-Paule and Nathalie alone to face her potential medical problems. Even with the transfusion, Molly was still seriously anemic. With her history—being born under the foundations of Miss June's little house, sharing the space not only with her mother and all the other puppies, but probably also with the cats, chickens, and the Lord alone knew what else—I knew she had to be riddled with intestinal parasites of just about every descriptions: tapeworms, whipworms, and worst of all, ascarids, whose larvae cause untold damage as they puncture through the wall of the intestine to enter the lymphatic system. Once there, the larvae migrate to the lungs, moult into the next stage of larval development, then burst through the delicate walls of the alveoli, the lungs' honeycomb network of microscopic air sacs. They are then coughed up and swallowed, only to complete their life cycle in the intestine, where they become egg-laying adults. I was afraid to treat Molly for these parasites, however, because of her very young age and dangerously weak condition. I feared the toxic side effects of the medications and wanted to wait awhile until she put a bit more meat on her bony, potbellied frame. Yet, by withholding treatment, I was risking Molly's condition becoming even worse.

Molly's eyes were also a great concern to me. They were both still badly infected, but I was particularly worried about her left eye, which was bulging and had a dull bluish gray cast to its surface. I had finally been able to separate the lids, and we were now able to treat her eyes with antibiotic ointment several times a day. Even so, I thought it was touch and go whether we would be able get on top of the infection and save the eye.

Four days after Terry and I had left for Africa, Molly's condition began to deteriorate rapidly. Even though Marie-Paule and Nathalie had continued to treat Molly's left eye with antibiotic ointment, the eye began to swell alarmingly and the puppy became anorexic, refusing all food.

Marie-Paule had tried to reach me by telephone in the Gambia, but Abay, the man who took care of Janis's house near Banjul, kept repeating in halting French, "He's upriver." The only way to get messages through to people "upriver" or "up country" in the Gambia at that time was to stand on the outskirts of Banjul, on the edge of the wide, dusty road, and try to wave down a passing bush taxi. If the taxi didn't break down, the driver would deliver the message to someone in the nearest town or village, and that someone would eventually get around to delivering it to its final destination. Needless to say, Marie-Paule decided to turn elsewhere for help.

In desperation, she telephoned Doug Cohn, my veterinary colleague at the laboratory, to seek his advice and help. Dr. Cohn was on weekend duty and was checking on the animals at the lab. LEMSIP had 250 chimpanzees and about 300 monkeys at the time, an often overwhelming responsibility when either one of us was left alone. But despite his workload, Doug told Marie-Paule and Nathie to bring Molly to the lab right away; he would see what he could do.

With Molly wrapped in a woolen blanket, Marie-Paule and Nathie drove to the lab. Doug examined Molly thoroughly and then put atropine drops into her eyes to dilate the pupils, so that he could examine the lenses and the retinas (the membranes at the back of each eye) with an ophthalmoscope. Doug was very concerned about Molly's left eye and wondered whether it might not have to be removed. He feared that infection might travel along the optic nerve, if it had not already done so, and affect her brain; she might then die from meningitis. While Doug sat in his office and consulted his surgical anatomy books for ideas on how best to go about treating Molly's condition, Nathalie took the puppy for a walk through the woods. Then disaster struck. Molly, trying to wipe her left eye free of the atropine drops that Doug had instilled, tore the cornea, the sensitive surface that covers the eye, with her dewclaw. Letting out a high-pitched scream as the transparent liquid of the anterior chamber of her eye exploded down the side of her face, Molly jumped backwards in pain, slammed into a tree, and collapsed unconscious on the ground.

Nathie recalls that she also screamed as she bent down to lift Molly up in her arms and ran back to the lab to alert Dr. Cohn. By now, Nathie was almost hysterical, certain that Molly was dying.

Doug quickly set about inserting an intravenous line to administer electrolyte solution and drugs to try and bring her out of her coma. "If we can pull her through her immediate crisis," Doug explained to Marie-Paule, "I'd urge you to have Molly's eye removed. I have a friend who's a veterinarian in Connecticut and a specialist in this type of surgery. I can call her now and see if she can set something up right away. I'll even go with you and Nathie to Connecticut, if you like."

Doug put a call through to his friend, explaining Molly's situation. He then put Marie-Paule on the line. The veterinarian agreed to perform surgery, but wouldn't be able to do it until the next day. On finding out that Molly was a Jamaican bush dog, however, the veterinarian became apprehensive, afraid that Molly might be carrying some rare tropical disease. "It might be better to bring your puppy at the end of the day after my other clients have left," she told Marie-Paule.

After an agonizing wait of an hour and a half, Molly began to regain consciousness, but she was still weak and tottery when she finally stood up. Doug had saved her life —even if only for the moment.

Marie-Paule now faced a dilemma. She didn't for one moment doubt that Molly might die if she didn't have surgery, yet she felt equally certain that Molly wouldn't survive the rigors of general anesthesia, not to mention the surgical procedure itself. On top of everything, Marie-Paule realized that she would have to deprive Molly of food for several hours before the surgery. Even though Molly had refused all food in the last many hours, Marie-Paule felt that further, purposeful deprivation of food could only harm her. Besides, the strain of the two-hour trip to Connecticut wouldn't be good for Molly, either. If Molly were to die, Marie-Paule felt, it was best to let her die at home, cradled by someone she loved and trusted, not in some cold, sterile veterinary clinic, prodded and poked by a stranger. Molly had been through too much anguish in her short life to die so miserably. Fully aware of the weight of her decision, Marie-Paule told the vet over the phone that she would chance it and not have the surgery done.

When I think back on it now, I find it hard to imagine the strain Marie-Paule must have been under, wondering if she had made the right decision, and what my opinion might have been. She and Nathalie took Molly home, both certain that she would die. Molly continued steadfastly to refuse all food, clamping her jaws together with every attempt either of them made to force-feed her. Marie-Paule was sure that

Molly would weaken to the point of death. That night, Nathie took Molly to bed with her, cradled in her arms, and attempted to force-feed her a beef liver broth she had prepared. With a one-milliliter syringe she repeatedly tried to force minute drops of broth into the side of Molly's mouth or between her clenched teeth, but to no avail.

In the middle of the night, Marie-Paule tried once more to get hold of me by telephone in the Gambia, but again Abay said, "He's still upriver."

Around five o'clock in the morning, after almost nonstop efforts, Nathie made a breakthrough: Molly began to lap the liver broth with vigor. Soon, Molly's condition had improved enormously. By then, she and Nathalie had established a deep bond that survives, undiminished, to this day.

Meanwhile, unaware of the drama back home, I proceeded with the surgery on Johnny. Terry and I realized at the moment we opened Johnny's abdomen that he had an incurable condition. The stomach wall, normally no more than a quarter of an inch across, was two to four inches thick, the lumen compressed to a mere slit. An infiltration of tough, pearly-white material, which separated the muscle layers from the delicate glandular tissue lining the stomach wall, extended the whole way up his esophagus, not a single inch of which was unaffected. He had sarcoidosis. The wonder was how had Johnny survived so long, in such remarkably good condition, when he had developed such severe pathology. We had no alternative but to euthanize him on the surgery table.

When I did finally get back to Banjul, after the operation on Johnny, I called Marie-Paule and learned what had happened. As has been the case so often over the years, I found myself away from home during a crisis. When I was in India, attending an international conference and training medical technicians, Marie-Paule had had to dig herself out from a monumental snowstorm. While I was on a trip to Peru, studying tamarins in the jungle and collecting semen from llamas in the high Andes, the basement of the house flooded. This time was much worse, however—I hadn't been there when Marie-Paule so needed my moral support.

Within four days of my return from the Gambia, I was off once again, on a previously arranged visit to California to take part in a daylong debate at an animal rights' convention in Sacramento on the morality and ethics of using animals in biomedical research. It was what you might call a very one-sided debate; I was the only one representing the "bad" guys, the scientific community, and I faced a united front of perhaps eighty people ready to tear me to shreds and deposit my bleeding remains in the nearest back alley. I couldn't help thinking that they had chosen the very worst person to debate, for I agreed with a good 95 percent of what they believed in, even to the extent of recognizing that animals do have rights, and we, as human beings, have no right but only a pressing need to experiment on them. I also recognized that without the pressure of the animal rights movement, the scientific community, by and large, would have made little effort to improve the lot of animals in the laboratory; after all, they hadn't done so until recently, and I saw no reason to suspect that they would have even then, left to their own volition. Just one thin but strong thread remained to be broken in my argument: the conviction that without using animals in research we wouldn't be able to make the advances in medical knowledge that have greatly improved human health. To my surprise, and immense relief, the debate turned out to be an exhilarating experience, and I came away, I like to think, having made some very good friends.

I returned home at around one-thirty in the morning and fell asleep as soon as my head hit the pillow. I was awakened with a start by Nathie's calling out, "Daddy!" It was three-thirty in the morning, pitch-black outside, and I had been asleep for only two hours. Marie-Paule and I jumped out of bed, rushing to Nathie's bedroom. There stood Nathie, tears streaming down her face, holding a rigid little Molly spread upside down across her open arms, Molly's four tiny legs sticking up stiffly in the air. An overpowering smell of fetid diarrhea emanated from the room.

"I think Molly's having a fit," Nathie cried, her words choked between great sucking sobs. "I woke up to this awful smell, and suddenly I heard Molly making these terrible screaming noises. When I turned on the light, she was rolling on her side with her legs cycling in spasms and her neck arched backwards." Already, even within these few seconds, Molly had become unconscious and now lay limp in Nathie's arms, without a sign of movement. The floor was covered in a great elongated streak of diarrhea.

I took Molly from Nathie's arms, carried her into Christopher's unoccupied bedroom, and lay her on the bed. Marie-Paule had already rushed downstairs to get my emergency medical box and stethoscope. Molly was in deep coma. Her eyes were rolled up into her head, the whites showing grotesquely, her breathing raspy and shallow. I placed the stethoscope on her chest. Her heart rate was so slow that I could hear the sounds of the individual valves of her heart opening and closing—a liquid, sloshing *floob-doob-doob, floob-doob-doob*—about 30 beats per minute, I guessed, instead of the almost uncountable 120 beats or so of a normal dog's heart. I felt her abdomen with both hands and was struck by the spasm of her contracted intestine—I could feel each individual loop, from one end to the other, like a tight, narrow cord of hard rubber. Her tongue and gums were white, and she was ice-cold to the touch.

Frantically, I grabbed a vial of atropine from the emergency box, the only drug I had that I thought could increase the strength of her heart, which was dangerously close to going into arrest. I quickly drew up half a milliliter into a syringe and needle and injected it deeply into Molly's thigh muscles. She was in shock, there was no doubt about it, but my mind was so foggy, I couldn't think straight enough to determine what might have caused it.

I showed Nathie how to rhythmically compress and release Molly's chest to assist her breathing while I filled another syringe with a steroid I hoped would stimulate blood flow to Molly's brain. Think, think, think! I kept saying as I struggled to jolt myself from the effects of the tiring return from California and the attendant jet lag. Remember your basic physiology: What could cause Molly to have a seizure with such rapid and profound effects? My immediate thought was a sudden massive hemorrhage of the brain, caused perhaps by a wandering parasite or some damage to her brain cells during gestation or soon after birth. I wouldn't be able to treat her if it was either of these. Then I counted off, in my mind, other conditions that could result in such catastrophic consequences. A hereditary defect? Not likely, I concluded, not in an outbred dog like Molly. What about vitamin deficiencies? I wondered, but seizures from such a cause are rarely so severe as to result in coma. Of course, there was always the good old standby of "idiopathic causes," a term that physicians and vets use when they haven't the faintest clue of the cause. No, the most likely cause is parasites, I told myself, a whole mass of them, ascarids most likely, which had blocked Molly's intestine at some narrow point. The acute, intense pain of having her intestine stretched by the mass of parasites would overwhelm her autonomic nervous system, plunging her into seizure and coma, as if a sudden surge of electricity had blown all her fuses.

I listened to Molly's chest again with the stethoscope. Her heart rate hadn't improved at all. Her eyes were still rolled up into her head, and when I raised the eyelid of her one good eye and shone a bright light into it, the pupil didn't respond, remaining widely dilated. I grabbed more atropine, double the dose this time, and injected it into Molly's leg. Between them, Nathie and Marie-Paule began to draw up sixty milliliters of dextrose solution into a syringe, in another attempt to counteract the shock, but I couldn't raise a vein, so low was Molly's blood pressure, and I had to make do with injecting the solution under the skin of her back.

I listened again to Molly's heartbeat; it was still alarmingly slow. It was only a minute or so since the last injection, but, if the atropine was going to work at all, it should have done so within seconds. While Nathie continued giving the sugar solution under the skin, I gave Molly a third injection of atropine, this time a slug larger than I would use on an adult chimpanzee. Immediately, her heart rate began to increase, and over the course of half a minute or so, it returned to normal. Yet Molly was still limp, her eyes rolled up into her head, fixed and dilated. I pinched the web of skin between her toes, but I could elicit no flinching reflex. Her eyes didn't blink when I touched the corneas, which are usually exquisitely sensitive.

"Molly is almost certainly brain damaged," I gently told Marie-Paule and Nathie. "There's nothing more I can do. She'll either come out of it on her own, or she'll die."

We stood looking down at Molly's little body lying limp

on the bed. I had never felt so helpless in all my career as a vet. This little dog had been through so much in her short two months of life, all the struggle and the pain—now to end like this. She had survived the fleas, the blood transfusion, had withstood the grueling journey from Jamaica to America, had gone through the pain of losing an eye, all for naught.

Nathie began to cry again, but this time silently, huge tears running down her cheeks. Marie-Paule stood without a word, her hands on my and Nathie's shoulders in comfort. The minutes went by into half an hour without change. Suddenly, Molly flinched. Was it my imagination? Was I willing myself to see something that wasn't there? No! There! Again, another flinch of the legs! With that, Molly sat bolt upright, sneezed, licked her chops, and wagged her tail enthusiastically. "Molly," we all cried and cuddled her into our arms. The little mite was back.

Chapter Fifteen

■ ■ ■

ANYONE WHO HAS EVER had a young puppy will know the sheer delight of seeing it develop and take on a character all its own. Molly was only four and a half weeks old when we got her to America, much younger than a puppy would normally be when weaned from the mother. Even a healthy puppy, at this young age, requires special care; for Molly, the prospects of her leading a long and uneventful life seemed slender indeed. Molly's almost continuous health problems—first her eye bursting, then her massive seizure, and always the constant fear of other things going wrong with her—hung over us like a dark cloud.

Despite several months of repeated treatments, Molly still had the bloated belly characteristic of a major parasite infection. Whether this was because of a particularly resistant parasite, or because she had developed a low grade, chronic infection of the intestine as a sequela to the parasites, I could not determine. We soon became able to predict her next crisis. Each time, her tummy would blow up like a balloon. Then we would be in for two days of diarrhea and uncontrollable vomiting, Molly unable to keep down even the smallest amount of food. I visualized the masses of two- to three-inch-long roundworms (like wads of white spaghetti) blocking the narrow parts of her intestinal tract in the duodenum or entrance to the colon. I imagined, too, the swollen walls of her delicate intestine, inflamed from the constant invasion by the bloodsucking worms.

Yet, after each trial, Molly bounced back with tenacity and spirit. She wasn't going to let anything hold her back. Except for a few photographs that I took of her in her early days, the only other record we have of her growing up is a videotape I had taken at each significant stage of her development. When we look back on it now, it's hard for us to believe just how tiny she once was, how tottery and weak.

One particular photograph, which I took within a day or two of bringing her home, shows Molly lying fast asleep next to one of my tennis shoes. Her little head, from the angle of her jaw to the crown, was only half the height of the heel of the shoe. The little woolly unicorn that Nathie gave her to play with when she first arrived in the house stood taller than she was.

When I think of this toy unicorn, I remember Finnegan and his little red teddy, the same one he clutched as his pet carrier headed down the airport conveyer belt. Nathie had given it to him when I had first brought him home from the lab, something soft for him to cuddle in his incubator during the night. Finnegan became very attached to his teddy, carrying it around with him as he explored the house. Sometimes we would find Finnegan sitting in some corner, having a soft, whimpering little conversation with Teddy, while he held him by the neck and stared into his eyes. But just as

Finnegan had felt duty bound to remind Christopher every evening who was boss, so he would beat the daylights out of Teddy each morning when he first woke up, while he waited in the incubator for us to prepare his breakfast milk and baby cereal. After Finn grew up, he no longer played with Teddy; I have it sitting on the bookshelf in my office now, a constant reminder of those far-off days. It is, surprisingly, as intact as it was the day Nathie first gave it to him.

As Molly grew, she began slowly to change in appearance, not just in shape, which you would expect, but in color, too. As a puppy of just a few weeks of age, she had well-defined patches of black over her eyes and down one side of the neck, and the one odd-looking little splotch of black on her bottom. The rest of her was pure white, except for the brown trim to her cheeks and the nutmeg spots above her eyes, the only visible traits that she had inherited from her father, Shaba. As the weeks went by, however, the pink skin of her tummy developed irregular black and blue blotches all over, and blue spots began to show on her flanks, then her legs, and finally her back. Molly seemed to be growing up in reverse: she was becoming more pigmented with age, unlike most animals—think of fawns or lion cubs, or even children, for that matter, with their childhood freckles—all of whom tend to lose their spots as they grow. With each passing day, Molly was looking more and more like her mother. We came to realize that, beneath the surface, Molly is more black than she is white; only the tips of most of her hairs are white, a feature that becomes very evident when we give her a bath.

One of our great pleasures was to see Molly take her first steps to explore the world around her. A milestone in any puppy's growing up is its overcoming the fear of climbing a flight of stairs, and the even more terrifying prospect of having to come down them again. And so it was with Molly: Nathie stayed close by her side, doting mother that she was, ready to grab Molly if she stumbled or fell, as she first climbed up, then down the staircase, one step at a time. Once successful, Molly danced and yapped with unbridled joy, eager to repeat the process.

I see this behavior with the chimps and monkeys at the lab. The moment comes when an infant first plucks up the courage to leave its mother's arms to explore the cage around it. Almost invariably, the baby first climbs the bars of the cage wall. The good mother, sensing that the infant is going to panic when it looks down and realizes just how high it has climbed, will sit waiting, arms outstretched, hands open, ready to take hold of the infant at the first sign of fear and guide it gently down to her breast and the security of her embrace.

Equally exciting is catching a puppy's first high-pitched, yappy little bark, as we had the fortune to do with Molly. Mystified, she looked around to see who had made this noise, only to find, to her surprise, that it was she. Overwhelmed by cockiness and the sense of achievement, she barked and barked with delight at anything that moved, like the fallen leaf that rustled in the breeze, or the seed-laden heads of grass swaying to and fro. These are indeed the moments to savor.

As, bit by bit, Molly overcame her health problems and began to grow, the characteristics of the wild dog in her started to emerge. For instance, when she eats from her bowl, Molly shakes every mouthful with rapid, almost violent flicks of her head, as if the food were alive and she is trying to break its neck. It's like the behavior of a lion who has felled a gazelle. This makes for quite a mess on the floor, as she splatters gravy from side to side, leaving a dripping trail as she carries each mouthful from her food dish back to her mat—her eating grounds—five or six feet away. What's so special about the mat? I wonder. Why does she feel obliged to transfer her food to it before she can eat? Sometimes, Molly's habit is a major annoyance. If Marie-Paule should decide to take the mat away to be washed, Molly will sneak her food downstairs, one mouthful at a time, to eat it on the brand-new carpet we recently installed in the basement.

Another way in which the wild dog in Molly is apparent to us has to do with her dewclaw. In domesticated dogs, the dewclaw has become vestigial, a lifeless digit that hangs limp on the inner side of the foreleg, just above the paw. The claw itself continues to grow, as if it belonged to a functional appendage, and it can become so curled and overgrown that it creates a hazard for the dog, becoming snagged and torn in branches or fences, or growing into the skin to create deep sores. For this reason, owners often have their dogs' dewclaws amputated. Molly's dewclaws are anything but vestigial. She has an unusually strong ability to flex and extend her toes and claws, and uses her dewclaws like thumbs. This gives her the ability to wrap her front paws around her feeding dish, which she sometimes even "carries," dragging it over short distances. It also enables her to secure big, juicy bones with a grasp as powerful as a feline paw or a primate hand. This dexterity is most striking when she reaches up to greet me when I return home from work. She affectionately grasps my wrist or elbow to pull me down to her level, and then places her "hands" on top of my head, her dewclaws clasping each side of my face to pinion me while she slobbers me with welcoming kisses. This, I am convinced, is a wild trait, an ability that might be essential to any hunting dog of the African savanna.

I told Roger Caras about Molly's dewclaws that night I met him in the New York City restaurant to discuss the origins of dogs. He seemed genuinely intrigued, assuring me he had never heard of a dog who could use its dewclaws.

"Does she expose her anus?" he asked me.

What on earth is he talking about? I thought, as I glanced around at the other diners to see if they might have overheard his lewd inquiry.

Undaunted, Roger went on to say, "The reason I ask is that wolves, unlike dogs, don't expose their anuses when other wolves come up to sniff them, because their scent glands are on the upper surface of the root of the tail, not around the anus, as they are in domesticated dogs. I'm just wondering whether Molly might show some original wolf traits. When she walks through the snow or mud, does she leave four footprints, like any normal dog, or only two, like the wolf, whose hind feet step into the prints of the forefeet?" Apparently this difference has to do with the different shape of the chest and how the shoulder blades hang down at the side. There was no wolf in Molly, I was sure of that, but she was a bush dog, nonetheless.

With the unexpected appearance of a trespasser on her land, Molly is transformed from the smooth-haired, butter-wouldn't-melt-in-her-mouth dog she normally is to a ferocious hyena. Her ears spring up, flecks of spit explode from her mouth, forepaws stab the ground as she jumps up and down on her hind legs, and the whole contour of her body changes, due to the coarse mane that, when raised, extends from the back of her head to the base of her tail. It's all show; if the intruder, which might be a Canada goose just landed by the pond, turns in its tracks, Molly hightails it, yelping, to the safety of the house and the reassurance of Mama. Molly is the only dog I have ever heard of to be chased by a doe, who I could only assume was trying to protect a fawn, hidden somewhere in the woods.

One peculiar trait, which I'm not sure has any connection to the wild, is Molly's ability to walk almost on what seems to be tiptoes. On those rare occasions when I'm the first one in the house to get up in the morning, I try not to wake anyone else as I step quietly down the stairs, careful to avoid the creaking steps in the middle of the flight. I'm just as cautious with the refrigerator when I go to get the milk, bread, butter, and jam that I'll need for breakfast, for fear that I'll awaken everyone with the sucking sound on opening the door, or the sloughing sound as it slams closed. I avoid clattering the plate or cutlery as I remove them from the cupboard, and I try as hard as I can not to grate the toaster as I load it up with bread.

Molly stays behind in bed during all of this, lying close by Marie-Paule's side. She barely wags her tail at me to say good morning when I first get out of bed, and you'd think she was so exhausted from the day before that she just has to have an extra wink or two before she can stir. It's all show, however, because I know she's listening intently for the noise of toast popping. This is the signal to her that, not only have I finished my cornflakes—and she had better be careful that Girlie, the cat, doesn't drink all the milk I have left in the dish for her and Molly to share—but I've also prepared the two slices of toast and jam that we share, alternating a piece at a time, just as I used to do with Spike, and then with Finnegan and Erin, all those years ago during our lunchtimes in the woods.

As I sit quietly at the table, still muzzy from the night's

sleep, I catch Molly out the corner of my eye, carefully tiptoeing down the last few steps of the staircase and across the floor of the kitchen to sit by my side. She walks as if her body were partially suspended by invisible ropes, only the tips of her paws coming in contact with the floor. On occasion she can be so deft and quiet that I don't hear her at all, and I'm surprised to find her sitting beside my chair. This behavior I find interesting in two respects. For one, how is she able, physically, to tiptoe? I try to catch her unawares, to study more intently how she does it, but as soon as she realizes I'm watching her, she reverts to her normal way of walking. Roger Caras didn't believe me at all when I related Molly's habit to him. He thinks it's physically impossible for a dog to tiptoe, and he may well be right. But what puzzles me even more than the question of whether a dog can tiptoe is why Molly walks this way in the first place. Is she, like me, trying not to wake up Mama when she leaves the bed and climbs down the stairs? Or does she think I want peace and quiet and she doesn't want to disturb my reverie? I don't think it's the latter, because if Marie-Paule is the first to get up, Molly doesn't hesitate to make noise and wake me up as she treads over my body, rushes downstairs like a bullet, and rockets out the front door to urinate and bark at the neighborhood. Either way, it's a fascinating behavior that begs explanation. Is it a demonstration that she's capable of empathy, of putting herself in another's position and not just seeing the world around her from her own perspective? Or is it indeed a wild trait, stemming from some sort of pack behavior, when individual dogs must act in careful concert with one another so as not to undermine the chances of a successful hunt before the final rush to the kill?

■ ■ ■

Sharing food with animals is an important way of establishing a relationship with them, as all pet owners know. Of course, the act of sharing doesn't necessarily imply a two-way street. It's almost always the human who's going to do the sharing, and the animal who's going to do the taking. Chimpanzees are the only animals I know of who truly share food, not only amongst themselves, but with human beings as well. I discovered this one evening not long after I had started working at LEMSIP, when I was making my evening rounds. I like to do my rounds at the end of the day, when most of the people at the lab have gone home. I know I'm not going to be bothered by the paging system, or be disturbed by the technicians entering the animal rooms. The chimps are at their quietest then, getting ready to bed down for the night, lying on their chain-suspended, truck-tire perches, which sway with their every movement. Even after long years in captivity, our chimps follow the same rhythm as wild chimps when they go to sleep in the branch-laden nests they construct each night, after sundown, at the tops of tall trees. All is calm during the hour before lights-out—you hear the chimps crunching lazily on their food, or giving out an occasional yawn, or making "raspberry" sounds as they force air through their tightly pursed lips while they groom themselves or one another—all sounds of quiet contentment.

On one particular evening, I came upon Jojo-M, an adult male chimp with an especially friendly disposition. He was sitting at the front of his cage, holding the bars with both hands while he munched on a hard biscuit (these biscuits, along with fruits and vegetables, constitute the basic diet we feed the chimps). Without putting too much thought into it, I said, "That looks like a very nice biscuit, Jojo. May I have one?" To my surprise, Jojo took the biscuit from his mouth and handed it me, insisting with a repeated nodding of his head that I try it. I pretended to eat off the corner of the biscuit, making great chomping sounds while doing so. Handing it back to him some moments later, I said in enthusiastic tones, "Thank you, Jojo. That was a very nice biscuit." He became quite excited by my response, bobbing his head up and down and bouncing on his bottom. Suddenly he made the most awful retching motion from the back of his throat, as though he were about to vomit from the very depths of his stomach. He brought forth into the palm of his hand a wad of finely chewed biscuit that he had carefully prepared by mixing small mouthfuls of biscuit with mouthfuls of water. Chimps love to do this with their biscuits, especially with their evening meal. Unlike their early morning meal, which they tend to scarf down in frenzied hunger, it can take them an hour or more in the evening to blend the biscuits into the perfect consistency. Some of them will actually prepare their entire biscuit meal in this way before swallowing even a crumb, storing each mouthful they have disgorged on ledges inside the cage, or on the crossbars of the walls, careful not to allow any to fall through the floor out of their reach.

As Jojo handed me this disgusting-looking sludge—it was the color and consistency of a semidried glob of diarrhea—he bounced up and down with a big panting grin, as if to say, "If you thought that biscuit was good, just wait till you taste this!" Again, so as not to offend him, I pretended to eat Jojo's gooey mess as it oozed between my fingers.

One of the most consistent food sharers among chimps I ever knew was Rufe, a venerable old gentleman who used to pass food over his shoulder to his females as they waited patiently behind him. The oddest example I ever saw of food-sharing was with Nancy and Stella, two grumpy old chimps who lived together with Rufe during the birth and rearing of two consecutive sets of babies. The first time round, Nancy and Stella had become pregnant, and then given birth, more or less at the same time, to female infants named Katherine and Pammy. Like most infant chimps, Katherine and Pammy began to experiment with eating solid food only once they reached about nine months of age. Even then, they used the biscuits more as playthings than as serious nourishment. Stella came up with a novel approach to encourage the infants to begin eating solid food. She would chew all her biscuits into the consistency of soft meatballs, just like Jojo-M, while she lay stretched out on her abdomen on the floor of the cage. Once her chewing had brought the food to its final consistency, Katherine and Pammy would approach, one at a time, squat in front of Stella's face, and, reaching forward with both hands, grasp her ears, touching their lips to hers as she extruded the stodgy mass into their mouths, like toothpaste being squeezed from a tube. This became Stella's standard way of feeding the two infants: Nancy seemed not the least bit disturbed by Stella's meddling with her infant. In fact, Nancy seemed quite content to have someone else assume her responsibilities. The second time around, with the birth of their next babies four years later, Nancy and Stella reversed roles, Nancy taking sole responsibility for feeding the infants the biscuit mash.

For all our closeness, Spike never shared food with me, except when it came to peaches. Peaches were, without doubt, his most favorite food. Just to see one would make him turn into a quivering jelly of excitement and expectation. His love for peaches began when he was a little fellow, living at home with us. I would take him in my arms to explore the two small peach trees behind the house, and he would reach up to pluck the fruits himself. Spike would always insist I take the first bite, and then would continue to offer me alternate bites until there was nothing left but the pit.

I have never known a dog to share food, except a mother with her puppies, and most dogs would snap if you attempted to take food from their mouths. Angus was certainly like that, snarling viciously if we pretended to go for a bone he was gnawing. Whether he would have actually bitten us if we had persisted, I don't know, because none of us wanted to put him to the test. Incredibly, Angus used to allow Finnegan to take the most brazen liberties with his food. The small monkey would stand over Angus's dish, daring him to do something about it, and would actually reach up and tweak Angus's nose if he didn't back down. For all his menacing growling, Angus never once attempted to bite Finnegan or put him in his place. Of course, Finnegan wasn't the least bit interested in Angus's food; it was just one more way to goad Angus on.

BY THE TIME MOLLY had reached six or seven months of age, she began to exhibit a most odd behavior. She discovered that one of the most relaxing positions of rest involves placing her bottom on one step of a staircase while her front paws are planted on the step beneath, as if the step she's sitting on were a little chair. She spends hours sometimes sitting in this pose; often, with her neck extended forward, her head drooped, she nods off to sleep, reminding me of a horse who has learned to sleep while still in the shafts. Every evening, Molly sits patiently on the top step of the back stairs, looking up the steeply inclined driveway through the end window of the house, listening intently, with her earflaps extended upwards, for the first sound of Marie-

Paule's return from school. I have no idea of the significance of her sitting stance; it seems to be just a unique quirk of Molly's, because I have never seen any other dog do it, wild or domesticated.

Apart from her penchant for garbage (she also sits in wait on the Tuesday and Friday pickup days for the crows who have a similar appetite), Molly is, on the whole, a very well-behaved dog with whom you can reason to quite a remarkable extent. I don't say this only as a proud parent might; I'll readily admit that Angus was a handful, constantly getting himself into trouble and becoming remarkably deaf when he decided he didn't want to toe the line. No, Molly is, as I say, a pretty well-behaved little dog. This is mostly because we don't have to give her specific commands to do, or not do, things. From a very young age, she has responded to conversational tones and is constantly listening, like her mother used to do, to what is being said, trying to pick up on key words so that she can stay ahead of the game. We have to be careful about certain words, like "whiskey," which happens to be the name of a springer spaniel that lives over the hill, whom she secretly admires, although she barks ferociously should he appear on our land.

This is not to say, however, that Molly doesn't sometimes behave badly. Then, we have a problem, as Molly becomes an indignant, resentful princess. Such crises are usually precipitated during our evening meal, when Molly may have picked up, with her incredibly acute sense of hearing, that some stranger has entered the little dead-end road that we live on, or that Whiskey has come to visit. She begins to bark and soon is out of control, despite my repeated scolding. "M-o-l-l-y," I say if she persists, drawing out the syllables for emphasis, "that's enough! Go sit on the mat!"

which she interprets to mean sit or lie anywhere on the floor. "Grmmmph," she replies in grumbling tone as she hunkers down to lie on the floor. "Molly," I repeat, but this time more sharply, "don't answer back."

"Hummph," Molly replies, in even more indignant tones as she squirms from side to side on the floor. If I persist beyond this point, I'm immediately sent to Coventry, and Molly totally ignores my presence, making up extra specially to Marie-Paule to emphasize her point. After a minute or so of this exchange, Molly can't take it any longer, and with a final series of grumbles she slinks off upstairs to bed. Sometimes Molly becomes so resentful that she won't come to lick my face when I finally get into bed, which is her usual way of saying good night. With the memory still fresh, she won't even greet me the following morning, nor come to the back door to see me off to work, as is her habit.

Marie-Paule gets very annoyed with me on these occasions. It's for me, she feels, to make the peace, not Molly. "After all," she chastises me, "Molly is only a little dog, and you're supposed to be the big 'veet,'" which is Marie-Paule's way of making fun of me by imitating the way the Scots farmers used to refer to me. I think it's the scientist in me, my desire to see how far one can go with an animal, to explore the limit of an animal's memory, its sense of self, its notion of right and wrong, or even whether it has a sense of humor, that makes me want to do it. It's a bit cruel, I suppose, but I can't resist the temptation.

Chimps, of course, are more intelligent than dogs. At work I quite often see them reacting to or participating in events in a way that leaves me in awe of their intellectual capabilities. But there is one particular incident that involved

Art, an old, former aerospace-research veteran, that I shall always remember.

Art is one of the biggest chimps we have at LEMSIP, a handsome mass of muscle and brawn, in spite of his receding gums and worn teeth. He isn't a nasty animal and doesn't show any dislike towards me—unlike Jaybee, who spits on me with real meanness every chance he gets, or Spot and Joanna, two females who disliked me from day one and try to gouge me with the crooked tip of a finger if I get too close to their cage. I, like all the technicians at the lab, mean nothing to Art. So long as he is fed, watered, and given whatever else he needs or demands, he cares not a fig for any of us. In a word, Art is arrogant.

One evening, during my rounds, I came to stand in front of Art's cage, which was situated midway down one side of the room. I looked up at him, sitting high above me on his tire perch, the characteristic expression of aloofness on his face. As I watched him gaze around the room, groom a little point on his elbow, and then scratch himself, totally ignoring my presence, I got to thinking that I had known this animal for fourteen years, and yet there was absolutely no understanding between us, no attempt at communication. I just didn't count in Art's world. I was determined to change this, and tonight was the night I would do it.

"How is he tonight, Art?" I called out, addressing him in the third person, as is my usual habit when talking to animals, and even very young children, now that I come to think about it. He ignored me. "Why won't that old Art speak to me?" I continued. With that, he lazily extended his left arm towards me and gave me a sharp flick of his wrist—that's chimpanzee for "Fuck off!" Not to be put off, as I had been so many times in the past, I persisted. "But, Art, I only want to talk to him," I said. Again, Art flicked his wrist at me, but more aggressively this time. I was crowding him, getting into his personal space, but I resolved to continue.

I reached up, and with one finger touched one of the bars on his cage. That was it: Art was really miffed now, and he began to strike out at me with with increasing frequency as I continued to prattle to him. So violent did his actions become that, on recoil, he inadvertently caught the corner of his left eye with a fingernail. It must have hurt like hell, because Art suddenly grasped his face in both hands, in obvious pain. It was at that moment I committed the unforgivable sin: I laughed at him, something you should never do to an animal, especially a chimp. With that, he released his left wrist, his right hand still grasping his injured eye, and blindly lashed out at me again and again. Things have got out of hand, I said to myself, and after seeing that Art had done no real damage to his eye, I decided I had better leave.

The next day, when I entered Art's room, he was sitting up on his tire, ignoring my presence. Recalling the previous evening's events and, noticing that Art's eye seemed perfectly fine now, I asked him, "How's his old eye, Art?" To my astonishment, Art responded by grabbing his left eye with one hand and flicking his other wrist at me violently.

But now, he and I have a relationship, and when I enter his room, I see him off in the distance, squinting through the bars of the cages to watch me coming. As I approach his cage and come to stand in front of him, he turns his gaze away from me, but now he has a smile on his face, and I can hear him making soft little laughing sounds to himself. Of course, we continue our game of "I don't care about you," but when I touch his cage now he pretends he doesn't want me to do this, and he presses a finger lightly on mine, just to remind me that it's his territory. It's very humbling to find yourself in the presence of an animal with such a sense of self, such haughtiness. His behavior on that morning showed me the extent of Art's memory, but even more incredible, his subsequent behavior showed his real sense of humor.

I've often wondered whether nonhuman animals respond only to events of the moment, or if they're able to project into the future. I'm not referring, of course, to animals like bears preparing for hibernation, when they gorge themselves to remain nourished until spring. I'm referring to whether animals, like human beings, can consciously plan ahead. The clearest example I ever had that a chimp might be capable of such planning was in an incident with Jaybee, the chimp who hates me with a passion. One evening—again during my rounds—I wandered in to talk to my old friend Jojo-M. As I sat before his cage chatting quietly to him, it occurred to me that Jaybee was in the cage directly behind me, on the opposite side of the room. That's amazing, I thought; Jaybee hasn't spat on me yet. I wondered if, as with Art, we might at long last have achieved some sort of friendship.

I cautiously swiveled around from my perch in front of Jojo's cage. To my surprise, Jaybee was sitting, with arms outstretched, holding on to the front bars of his cage. He had a relaxed expression—not the nasty, scrunched-up face I was accustomed to. "Hi, Jaybee," I said, hardly believing this turn of events. "How is he tonight?" He nodded his head repeatedly, a smile beginning to dawn on his face. "It's so nice to see the Jaybee," I continued, careful not to move from my crouched position so as not to startle him and break the spell. Jaybee continued to nod at me, obviously happy, or so I thought. He then turned his head sharply to stare intently at the small window in the door about twenty feet away at the end of the room. I, too, turned on my heels to look at the window, assuming Jaybee was responding to the presence of one of the technicians.

"But there's no one there," I said to Jaybee. He continued to nod his head and smile sweetly. This continued for a couple of minutes until Jaybee again stared at the window. This time I was certain I hadn't heard one of the technicians. "There's no one there, Jaybee," I assured him. We continued on these friendly terms for a bit longer until Jaybee, for a third time, turned his attention to the window. He knew as well as I did that no one was at the window; he was creating a diversion. As I had turned away from Jaybee every so often to pay attention to Jojo or to look at the window, Jaybee had, inch by inch, leaned sideways until he was in easy reach of his water container. Before I could rise from my crouched position, Jaybee had sucked up a huge mouthful of water—a half gallon or more—and spat it full force over me, drenching me to the skin.

OUR FAMILY HAS ALWAYS had the habit of talking to our animals, including them in just about every conversation, asking them questions about what they would like, or what they might want to do, and never dreaming of walking past one without at least some word of recognition. We started talking to Molly from a very early age; probably as a result, she is more responsive to vocal cues than any dog I have ever met.

It is through such constant talking that an animal can pick up a great deal of understanding of human language, and, most likely, the earlier in life this process begins, the greater the extent of understanding becomes. I think this is probably true for all animals, not just dogs, and may have little to do with species, or intelligence, or even the size of their brains. Even, or perhaps especially, for animals confined in laboratory cages, talking to them becomes essential. After all, they are deprived of so many basic rights—the freedom of movement, the freedom to pick and choose their food and even their friends. These animals are so totally at our mercy, the least we can do is recognize their existence by talking to them and treating them as individuals. This is something I especially try to instill in the caregivers and technicians who work with the primates at LEMSIP.

Yet, for all my encouraging, I find that the men at the lab are less inclined to be talkative with animals than the women are. I'm not sure why this is so, but it may be because men feel embarrassed. It may be even a reflection of their lack of maternalism; after all, it's usually the mother, not the father, who's mainly responsible for teaching language to a child. Whatever it is, the women at the lab are the most consistent in their talking to the animals, and the woman who takes first prize is Bobbie.

Bobbie hails from New Jersey, and has the accent unique to that state. Worms to her are *woyms,* oil is *earl,* and duty is *dudy.* She is a veritable chatterbox. The monkeys, especially the Java macaques, which she seems to understand best, adore her. Sometimes, when I'm in the nearby clinic, concentrating on some delicate procedure or listening to an animal's lungs with a stethoscope, she nearly drives me to distraction, blathering away off in the background. But the monkeys love it. As she goes about her morning work, bending down beneath the cages to clean the floor, or scrubbing the walls, or filling the feed barrels with biscuits, or attending to any other of the score of duties she has, Bobbie chatters to them: "I don't know about all you monkeys, you make such a mess!" or "What's up, Juan? You don't like your celery?" or "You're such a good girl, April, letting me tickle your tail like that." And so she goes, hour after hour, the monkeys enthralled. I can hear their plaintive "whoo" replies, which don't denote sadness, but rather a heightening curiosity, as they await the next treat Bobbie has in store for them. Then they all call out in their crisp little "whip-whip" sounds, their long-awaited anticipation about to be fulfilled. Despite this rapport, Bobbie still often finds her work difficult—as do many of the caregivers at the lab—because of the clash between their love for the animals and the demands of the research.

As much as possible, my other duties permitting, I try to keep in close contact with the animals. I've known many of the chimps at LEMSIP for close to twenty years now. I brought many of them into the world, and the technicians and I have sat up many a night struggling to get them through some particularly severe illness. Over the years, I've established some special communication with each one of them. When I enter his room, Conrad, a large, thirty-year-old male chimpanzee, immediately sticks his tongue out through the bars, because he knows I'll come over to grab it with my fingers and wiggle it up and down in rhythm to a little song I sing to him every time. Not a particularly melodious song, it goes something like:

Iddle-liddle-liddle-dee,
Iddle-liddle-liddle-doh,
Iddle-liddle-iddle-dee . . .

After two refrains, it ends on a sharply rising and increasingly drawn out note:

Diddle-diddle, diddle-diddle, diddle-diddle-DEE!

at which point Conrad screws up his face, because he knows I'm going to tweak his nose as I let go of his tongue.

I greet Libbel, a ten-year-old female chimpanzee, with the rhyme:

> There was an old lady who lived in a shoe,
> She had so M-A-N-Y CHIL-DREN
> She didn't [and here I speed up] KNOW WHAT TO DO!
> DID SHE?

At this, Libbel laughs, in her grunting, out-of-breath sort of way, and she performs a little pirouette. All of the animals wait for these special interactions, and I would no more dream of walking past one of them without stopping to say hello than of flying to the moon.

Chapter Sixteen

■ ■ ■

ONE OF OUR MAIN concerns about bringing Molly back to America was that she would have a lonely life with no other dogs her own age to play with. Of course, there were Girlie and Snowy, our two cats, but I couldn't imagine their being interested in a tiny puppy, knowing cats for the snobs they can be. We had adopted Girlie, a tabby, from the local humane society after she had been found abandoned in a brown paper bag outside its gates. Snowy, as her name would imply, is as white as the snow and had lived her whole young life in a fourth-floor apartment in upper Manhattan, never knowing what it was to walk on grass or chase birds through the bushes, left by her owner for weeks on end until Pádraig and Nathalie, who shared the apartment while studying for their master's degrees, more or less kidnapped her. Within no time Snowy became the undisputed queen of the house; even Angus had to learn to keep a respectful distance from her, and poor Girlie was displaced from all her favorite spots, including our bed at night. One cold, slightly strabismic look from Snowy was enough to put the fear of God into the other animals and keep them in their place. To my surprise, Girlie did show a certain motherliness towards Molly, but Snowy was absolutely hateful, staring Molly down with her evil eye, and Molly didn't dare come within ten feet of her. Initially, as I had expected, Asa and Baxter, our neighbors' dogs, had no patience for Molly at all and seemed almost embarrassed by her presence, not knowing how to respond whenever she approached them. This was to change in time, however, as Molly became older, and now they are all good friends.

Within a few weeks of Molly's arrival, in a fortunate turn of events, one set of neighbors acquired a Beagle puppy by the name of Nikki, who was one week older than Molly. Nikki and Molly became friends and would wait at each other's back doors for their playmate. As they grew older, their play became rougher, requiring more and more skill as they chased each other around in ever tighter circles, tripping and falling over each other to end up in a grappling, tooth-gnashing mass. This rough-and-tumble play, so important for all young animals, no matter what the species, not only helped to toughen Molly up, but also helped her learn how to compensate for her one-sided vision.

Molly's first winter in America couldn't have been more of a contrast from the tropical heat of Jamaica. It was one of the worst winters on record for New York State, with a total of twenty-one major snowstorms. Yet Molly didn't seem to mind the cold in the least. She didn't hesitate to run out into the snow; even when she would crash through the crusty surface, momentarily disappearing, she would bounce back up and carry on.

With increasing confidence in herself, Molly began as

she got older to explore the land around her. She seemed, early on, to recognize the actual limits of our one-acre plot, even though it's not marked by any fence or border. I've always been mystified by this sense of territory that animals seem to possess, for even the two little rhesus monkeys, Finnegan and Erin, recognized at a very early age the limits of their domain. When Finnegan was only five or six months old, he would patrol the perimeter, his tail erect, the tip kinked, the hair on the back of his head raised, his gait purposeful—the typical posture of the alpha male—ready to challenge anyone who came near. But he would never, himself, cross that invisible line.

I never cease to be amazed at how early in life animals develop this sense of territoriality. I started keeping a diary of major events concerning the birds and other wildlife around our house seventeen years ago. I record such happenings as the first appearance of the red-winged blackbirds; the migratory ducks, as they stop off briefly on the pond before resuming their journey north; the Baltimore orioles, with the coming of spring; and the first observation of juncos, the unwelcome harbingers of the dreaded winter to come. Every so often when I arrive home from work at night, I sit with a bottle of beer and go through the diary, to refresh my memory. I can't help laughing to myself when I read some of the long-forgotten comments, like the one I wrote about the pair of red-eyed vireos I saw fighting with a starling while, on the very same branch, only a few feet away, a robin fed its young, totally oblivious to the racket; or the evening a sharp-shinned hawk dove to catch a chipmunk, almost colliding with my knees.

But there was one note that I discovered only recently, recorded for October 13, 1980, which reads "Spike showed his first masculine trait by stamping his feet on [the] territorial borders [of the land]"; Spike was just shy of eleven months old at the time. I can't specifically recall this incident, yet I can imagine Spike in my mind's eye, a tiny bundle of black fur, standing on his knuckles, alternately pounding the earth with one foot and then the other, his lips almost certainly extended into a tubular shape as he announces to the world "Oogh, oogh"—"This is my land!"

As Molly began to establish a closer friendship with Baxter and Asa, she extended her territory to include six or seven acres of woodland and open grassy areas. Each morning, she goes on her rounds, to check the deer trails, sniffing out the deer's overnight sleeping areas, and she traces in wild circles and zigzag lines, nose pressed to the ground, the scent of the rabbits. Molly has developed the grace and speed of an antelope in flight, sailing over the rocks and careering beneath the low-hanging branches of the trees and bushes. My one big worry is that Molly will get carried away if she picks up the tracks of the black bear who occasionally shows up in our woods.

One of Molly's greatest pleasures is to rush up to the ducks, slowing only at the last second to give them a chance to escape into the pond in a frantic splash, their wings all a-flap. She is so proud of herself then, and stands on the edge of the water with her tongue hanging out, huffing and panting with a sound that closely resembles a chimpanzee's laugh. The ducks don't seem the least bit scared by her, for they immediately circle back to the embankment to cackle and quack at her.

When you have animals, or, in our case, when you're liable to bring home baby chimps or even little rhesus monkeys, the value of good neighbors can never be overrated. And we are blessed by having just about the greatest neighbors anyone could wish for. Not only does everyone respect one another's privacy, but they are there, at a moment's notice, ready to help out in times of crisis. I provide the odd emergency treatment for their animals (as when Baxter was hit on the road by a car in the middle of the night), Don helps us plough snow, Richard lends a hand in cutting up a fallen tree with his chainsaw, and Arnold is always there to give Marie-Paule the latest information on the stock market. And no one has ever complained about Angus's rovings, or Molly's intrusions.

Short, portly, and the spitting image of the mustachioed character who advertises Dunkin' Donuts on television, Arnold, our neighbor across the pond, retired some years ago from a career as a photographer on film sets. Except for a stint he did in England in the Air Force during the Second World War, Arnold has lived his whole life in New York City.

When he first arrived in our neck of the woods, Arnold knew nothing more about nature and animals in general, other than what he had gleaned from watching nature programs on public television. But with the zeal of the newly converted, he quickly developed the fervor for observing the wildlife around him.

For being barely more than fifty miles northwest of New York City, our area is surprisingly wild and unspoiled, ideal for anyone interested in nature. Arnold and his wife, Brenda, a nurse from Yorkshire, own five acres of land that border on our one acre, and we share a fairly large pond and the stream that feeds it. A herd of deer frequents the forest, as well as the expected contingents of raccoons, possums, and skunks. Large bass and bluegills and painted turtles fill the pond, and a family of muskrats has lived for years beneath the hump that borders the stream. In summer, the great blue and the little green herons line the water's edge, and the kingfisher makes its dive from the overhanging tree branches, its raucous laughing call echoing through the still morning air.

Shortly after he and Brenda moved in, Arnold found himself the proud new owner of a small brown and white goat that had been abandoned. He turned her loose in a half-acre paddock in the middle of his land that the former residents had used to graze a pony. This I found rather foolhardy for the horizontal rails of the enclosure were widely spaced apart, ample room for a small, inquisitive goat to climb through. Sure enough, at around ten o'clock that first night, Arnold phoned to say that the goat had escaped and was nowhere to be seen. "What am I going to do, Jim?" he asked in confusion. For want of a better idea, I suggested to Arnold that he stand there, in the dark, and bleat: perhaps the goat would respond to his call and come back.

Half an hour later, a greatly elated Arnold phoned back to say, "She came right back."

"What exactly did you do, Arnold?" I inquired.

"I did just what you told me I should do: I stood there and called out 'bleep, bleep.'"

"But, Arnold," I replied, "goats don't bleep, they bleat."

Arnold had never heard this word before. "Well, whatever," he said, "she came back." And so the goat was christened "Bleep."

But Arnold and Brenda's greatest love is Asa, their big, soft very intelligent Rhodesian ridgeback, who was rescued from the local humane society. She had been friends with Angus, although she tends to stand aloof from Baxter, the little cocker spaniel who lives next door to us. Marie-Paule often refers to Asa as the "vacuum cleaner."

Whenever Arnold lets Asa out, and Asa thinks Marie-

Paule might be home, she lumbers across the oak bridge I constructed over the stream, races up the stairs to the balcony, and with the full force of her massive weight lunges through the back door into the kitchen. She then sets about gobbling up any morsels of food that Girlie or Molly might have left in their dishes, and then makes a beeline for Molly's mat. Molly particularly likes to sit on her mat and gnaw on bones, chunks and splinters of which get caught up in the mat's fibers. It's these leftover scraps that Asa relishes. Asa stands full square on the mat and yanks it up in her mouth, using her teeth to sieve out the bone fragments. At the end, not a speck remains, and the mat is as clean as if it had just been vacuumed. Molly gets very agitated by this and halfheartedly tries to block Asa from getting to the mat, but she's no match for Asa's Clydesdale size.

Unlike Molly, who has extended her territory back into the woods, Nikki began to wander toward the main road. Unable to break her of this behavior, and fearful that she would one day be killed by a car or cause an accident, Nikki's owners parted with her, adopting her out to a farmer. Now, only Asa and Baxter remain as Molly's close friends, and no other dog is allowed on the land. She tolerates Asa and Baxter coming to the house, although she would never think of going to theirs, but she jealously watches them to make sure they don't take liberties with us by coming too close or sharing a little of our love and affection.

All in all, Molly has a very rich life, with lots of love, our initial fears that she would lead a lonely existence in America totally unfounded.

Chapter Seventeen

■ ■ ■

MARIE-PAULE AND I went back to Jamaica in the winter of 1995, to a different villa this time, but still within easy traveling distance of Treasure Beach and Black River.

Our Molly was then about eighteen months old, still a puppy at heart, but her adult personality was well on the way to being formed. She seemed finally to be over all her medical problems. Nathalie had returned from Spain for a brief while and offered to baby-sit "the puppy," as she still refers to Molly, while we were away.

I wanted to take the opportunity to revisit the scenes of Molly's early life, like any would-be writer of a book, trying to add final touches to the manuscript and check details for accuracy. In particular, I wanted to visit Miss June in Treasure Beach and see Molly's mother, go by the doctors' offices and the pharmacy in Black River. Above all, I wanted to tell Dr. Francis how grateful I was for his help in providing the transfusion supplies during my time of need, and how Molly had finally overcome all her problems, thanks largely to him.

In the afternoon of the last day of our vacation, we set out for Black River. Coming to the hospital, we drove in through the gates and parked in the shade of the wall by the men's ward. The walls were the same sun-bleached white, and the mosquito screens in the tall windows were still missing or flapping in torn strands in the breeze. A crowd was already beginning to gather around the fruit and jerk stands outside in the courtyard.

Marie-Paule waited in the car while I made my way to the single-story building that housed the outpatient department. It was as crowded and chaotic as I remembered, the walls the same greasy two-tone off-white and gray, but a pleasant cool breeze wafted through the immense louvered windows on either side. I approached the examination room, knocked lightly on the door, and noticed that the missing doorknob had still not been replaced.

A nurse came to the door, and I asked her whether Dr. Francis was on duty, and if I might be able to speak with him. Her smile faded, and her face took on an undeniable sadness. I felt my heart sink.

"I'm sorry to tell you, but Dr. Francis died last March," she said softly.

"That's terrible," I replied not, wanting to believe what I was hearing. "Whatever happened?"

"Oh, he was attending a dance in celebration of a fund-raising campaign he had organized for the high school—he was always trying to raise money for some worthy cause or other—and he just collapsed and died," the nurse informed me.

"Oh," I went on, "that is so terribly sad."

"Well," she said, a smile beginning to brighten her face," Dr. Francis always used to say, 'I hope when my time comes, it will be quick and I shall be doing something that I enjoy,' and he was certainly having a great time dancing and chatting with everyone," she said, chuckling. "You know, he was only seventy years old but he'd had a bad heart condition and had to retire from his full-time duties at the hospital. Everyone came to pay their respects, and you know, they gave him a state funeral down at the big church, he was so important and everybody loved him so." The nurse went on to recount Dr. Francis's many accomplishments—he was a preacher revered throughout the parish, he was a justice of the peace, he was a prominent leader in education, he'd been awarded the national Order of Distinction, and on and on she went, brimming with pride.

I walked back out of the outpatient building, saddened by the news of the loss of this kind man. Little had I realized when I had called on Dr. Francis that hot afternoon a year and a half before what a great man he was, so loved throughout the island for his kindness and medical skills. "He took out my appendix," one old man recalled when I mentioned Dr. Francis's name, and another informed me that Dr. Francis had saved his life when he developed a liver abscess. Dr. Francis, it seemed, filled a role in everyone's heart and imagination that was larger than life. Yet I recalled how close I had come to walking out the hospital that afternoon, trying to think of some alternative to solving my problem, afraid that Dr. Francis would not understand my concern for a helpless runt puppy, and might even become angry when he discovered the purpose of my taking up his time. Yet he hadn't been annoyed with me. Now I would never have the chance to chat with him, to thank him for the kindness he had shown me, and to try and find out what he had really thought of me and my story about the little Molly that needed his help.

Marie-Paule and I drove on into the little town, past the three doctors' offices, and parked the car by the supermarket in the center. While Marie-Paule went to do a little shopping, I set out on a nostalgic walk. The pharmacy was a little way up the main street, on the other side of the road. As before, two or three heavyset women in head scarves sat or squatted on the boardwalk outside the faded brown doors of the pharmacist's, selling their neatly arranged piles of fruits and vegetables. Inside, however, things were not quite as I remembered. The short row of wickerbacked wooden chairs was still there, where the little old ladies had sat in their lace gloves, waiting patiently for their prescriptions, but the glass counters had been rearranged, and the apothecary's jars were gone. Instead, the back of the shop had been converted into a small clothing department. The same solemn-looking lady was there, but this time she was attending to two ladies who were interested in buying a floral print dress.

On the surface, everything seemed the same, yet nothing was as I remembered it. As I walked back to the car, I felt dispirited and empty. The pharmacist's had changed, it no longer had the same old-world charm that I had remembered. Even the people who crowded the sidewalks didn't seem to have the same infectious, carefree effervescence that I had recalled, either. But perhaps it was me, and my somber mood, that were different. Dr. Francis had passed away, and I couldn't help feeling that this man would have had something to teach me about life. Born and raised in the community in which he had served in so many varied ways, I imagined he might have been a down-to-earth man with a simple philosophy, but I would never have the opportunity to find out now.

Marie-Paule and I next made our way to Treasure Beach to see Miss June. We had written to her from America, enclosing the photographs I had taken the previous year of the little children in her Bible class and Molly senior and her puppies.

We drove in through the open gates to Miss June's yard. She was delighted to see us, and gave us warm hugs and kisses. Her little house was the same, with neat piles of books stacked on tiny shelves here and there around the walls. Her nanny was with her, a bright, talkative ninety-five-year-old lady who looked to me no more than sixty-five.

I saw no puppies scampering around, although the hens and the surly-looking white cockerel were in evidence. "Where's Molly?" I asked Miss June. Miss June looked at me sadly and said, "Oh, Molly died last August, when I was away in Mandeville. We think she was poisoned, because two other dogs in the area died around the same time and with the same symptoms." I was dumbstruck. Molly had been only four or five years old, and she had had only three litters, the last just six or seven months ago.

"Yes, it was terrible," Miss June continued. "As I was about to leave in the taxi to go to Mandeville, Molly looked up at me with an all-knowing expression of sadness, as though she wanted to tell me something, and I had this awful premonition that we might not see each other again."

Miss June began to reminisce about Molly and her funny little ways. "Molly would hang on every word I spoke," she recalled, "waiting to catch some phrase or word that she knew and could react to." I was reminded of our own Molly, her mother's daughter for sure. "She loved to hear us sing hymns," Miss June continued, in her singsong Jamaican tone. "She would come and sit at my feet, and howl in such a melodious sort of way, with her little head stretched up in the air. She really was a most special dog!" That was certain. No more trips to Jake's, her gourmet restaurant, I thought, no more scampering masses of puppies.

"What about Shaba?" I asked Miss June, wondering about our Molly's father.

"Oh, he died quite some time ago," she replied. "Another dog came along and wanted to take over Shaba's territory. They had a terrible fight, and Shaba got so badly bitten, he died of his wounds."

I was shocked by the news. Life seemed so fragile, as wild and as savage as for any dog living in Africa. Our little Molly could never have survived these challenges.

Marie-Paule and I stayed for a little while to chat with Miss June and her nanny, but it was getting late and we finally left, both of us in sad and contemplative moods.

I decided to take the desolate coast road back to Black River, a road that had only recently been constructed, from crushed coral and seashells. Civilization had not yet caught up with it; no villas or hotels, not even any villages along the way, just miles of endless bush and rugged cliffs. The sunset was one of the most magnificent I had ever seen, the sun a fiery red ball hovering above the horizon, the sky streaked layer upon layer in lilacs, mauves, and pinks, the hues reflected in the smooth pale blue sea. The Santa Cruz Mountains, far off to the north, were aglow in the same soft pastels. It was an evening for reflection, and Marie-Paule and I sat silently as we drove, lost in our private thoughts as the night gathered around us.

In the great turning of this world, big Molly's passing was of little consequence, I supposed, yet I couldn't help feeling an awful emptiness. She was a very special dog, and the world was a little poorer for her leaving. Shaba had been a rough-and-ready sort of dog, nothing exactly lovable about him, and yet even he had served his purpose in nature, had put up the good, brave fight to ensure that his, and only his, genes got passed on to the next generation.

I couldn't help but think of my own life, my work, and its value. I had been in medical research for close to thirty years now, telling myself that what I did was for a noble cause, even though it gave me constant heartache to use animals so. If I were to stop tomorrow, would it make any real difference? Maybe I could avoid it all by finding a teaching job or going back into practice, anything to get away from the nagging in the back of my mind. But that would be cowardly of me.

Was our use of animals really essential? I knew well that animal rights advocates often lambaste scientists for the unnecessary repetition of experiments. They even go so far as to say that research on animals has not only failed to benefit man, but in some cases has led to erroneous results that have actually caused harm to human beings.

Yet it's no myth, I told myself, that without research on dogs, the very same sorts of dogs we all love as pets—many of them just like Molly—the treatment for diabetes couldn't have been found. And what about poliomyelitis? Most younger people in countries like the United States and in Western Europe are unaware of the horrors of infantile paralysis, as the disease was commonly known when I was young. Those who did not die from the acute effects of the disease often led the rest of their lives immobilized in iron lungs, because they could not breathe unaided, or wore heavy metal braces on their paralyzed legs. Without the rhesus monkey, vaccines against polio couldn't have been developed, albeit at the cost of hundreds of thousands of monkeys' lives. Then I thought of the chimpanzees, who play an essential role in the frustrating attempts to develop vaccines to protect against AIDS. Without the chimpanzees, we would have no vaccine against hepatitis B. There are over 200 million carriers of this disease worldwide, 40 percent of whom, in Third World countries, will die before the age of forty from liver cancer.

Did I really believe my own litany? Or was I like the dedicated young Communist I used to see when I was a student in Glasgow, repeating by rote the old party line to anyone who would listen as he stood frozen in the windswept street, handing out political leaflets?

I remembered once putting myself to the test before a mixed audience of scientists, animal technologists, and animal rights advocates. I had been invited to England to deliver the keynote address at an annual conference on the ethics of using chimpanzees in biomedical research. Eighteen years had passed since I had lived in the United Kingdom, and I was now hopelessly out of touch with British public opinion. Chimpanzees aren't used in biomedical research in Britain, that much I knew, and I wondered nervously what the audience's attitude would be to my using them at LEMSIP. Would they start throwing rotten tomatoes at me as I began my talk? Had they drawn the line at this species, determined that, because of some point of ethics or morality, they could never bring themselves to use an animal so closely related to humans? Yet, I also knew that there were no chimpanzees in laboratories in Britain that were used in research, in any event. "You don't miss what you ain't got," as the Cockney expression goes. They, themselves, did not face the ethical issue, at least, not in their own backyard. It is so easy for people who do not live in glass houses to throw stones at those who do.

I never write out a speech or practice beforehand what

I'm going to say. In fact, except for the opening line, which has to be strongly delivered, I usually haven't the faintest idea of what I'll say when the time comes to stand at the lectern; I leave it more to the audience's facial expressions, or body gestures—a raised eyebrow here, a quizzical look there, a hunching of the shoulders or disbelieving shake of the head—to guide me through, to mold and sculpt the topic to what seems to be generating the greatest interest. Without a set plan of attack in hand, I'm usually a nervous wreck by the time I reach the podium.

About two-thirds of the way through my presentation—the audience had not yet pelted me with putrescent fruit or interrupted me with catcalls—I thought of a question that I had never before posed to myself, the answer to which would prove to me, beyond any doubt, the extent of my own conviction. "Imagine this situation," I said tentatively, trying to work out quickly in my mind how I would make the story go. "A situation so hypothetical that for a thousand and one solid scientific reasons could never be, but let us pretend for the moment that it could be." I briefly explained to the audience about Spike, the chimp that I had delivered by cesarean section years before, and how, after his mother's death, I had taken him home to be reared by my family for the first year of his life, rather than leave him all alone in an incubator. I went on to describe how, until he was about eight years old, I would, on warm days, take him into the woods so that we could have lunch together. These excursions gave him the opportunity to feel the grass beneath his feet and climb the trees. I went on to explain how he and I had developed a very close bond. I loved Spike with a passion that I couldn't begin to put into words—not because he was like a pet, for I never looked upon him in that way, but because I was consumed by an overriding sense of responsibility towards him. I had saved his life as a baby, and without a mother to care for him, or other infants that he could grow up with, I owed him something special.

Spike represented for me the very epitome of the plight of the animal in research, deprived even in his young life of the experience that was his right to expect. "Now imagine," I continued, "that I find myself walking through the slums of Calcutta, and I see a child sitting in the gutter, emaciated, covered in sores, dressed in rags, and next to him is Spike. Imagine further that a world-eminent surgeon steps out from one of the hovels and tells me, in no uncertain terms, that if I don't immediately give up Spike so that the surgeon can take his heart or his liver for transplant, this little child will be dead by the following day.

"The question is," I continued, "could I bring myself to turn my back on that little child, a child whom I have never met before, and am unlikely ever to see again, or would I let the surgeon sacrifice Spike and spend the rest of my life in torment?" I almost froze on the stage when I realized the depth of the question I had just asked myself. "Yet, if I believe fully in what I espouse," I concluded, "I should be able to answer this question with a resounding yes: the life of one child is infinitely more precious than that of one chimpanzee." There was a deathly silence, broken eventually by a young veterinary surgeon from the Royal Society for the Prevention of Cruelty to Animals, who stood up to say that she absolutely disagreed with any research that called for only one human life saved for every animal destroyed. I agreed with her absolutely, but this was a different issue and she had missed the point I was trying to make.

As Marie-Paule and I continued on towards Black River

in the glow of the setting sun, I realized that, five years after that speech, I still hadn't been able to bring myself to answer the question.

Dr. Fred Prince of the New York Blood Center once wrote that the time between first isolating the virus that causes hepatitis B in 1968 and marketing the first effective vaccine for human use was a mere ten years, a truly remarkable achievement. It had taken, he estimated, fewer than 200 chimpanzees to reach this point, although, as no one can deny, a far greater number of chimpanzees have been used in subsequent vaccine development throughout the world. With over 200 million carriers of hepatitis B in the world, the ratio of chimpanzees used in the research to the number of human lives that could potentially be saved is huge indeed. Were all medical research so frugal and effective in using animals, we wouldn't face the weighty ethical issues we do today. Yet the burning issue is whether we have the right to use chimps, or any other animal, in the first place. If the answer is yes, then by extension, one might ask "Isn't Spike's life worth sacrificing, even if only to save a single human being? Who's going to be the judge?"

Every now and then, though, I come face to face with human suffering, and I'm taken out of the closed world of the laboratory, with its neatly laundered white lab coats, gleaming test tubes, and the latest piece of newfangled equipment. It's hard to maintain the goal in mind, this helping of humanity, when you aren't faced with the "end product"—the human beings in need. Then I thought of the hospital I had visited in Guinea, on my most recent trip to West Africa, just a month or so earlier. For a little while at least, what I saw there brought me back on track.

Guinea is the poorest country I have been to in Africa, although I'm sure there are poorer. On my journey there, I met an American woman named Pam, whose husband is the chief executive officer of a multinational bauxite mining company. She took me to a straggling, ramshackle little town in the northwestern part of the country, close to the border with Guinea-Bissau. The town had grown up around the smelting factory. Pam first showed me the fish market, a vast, sprawling sea of tin-roofed stalls, overflowing with bustling humanity, spread along the banks of a wide, lazy river. Then she took me to see a soccer game, some of the youthful players dressed in brand-new cleated boots (though most of the players were barefoot), the crowd of onlookers roaring with enthusiasm. Finally she asked me whether I would like to visit the local hospital, which was financed to a great extent by the mining company.

We entered the hospital through broad white gates and made our way among modest but well-tended gardens to the children's wards. Opening the door to the first ward, we found six adult-size beds. In the first bed to the left was a near-naked boy of about fourteen. His emaciated body lay twisted on the mattress, his mouth gaped rigidly open, his eyes rolled up into his head, only their whites showing. Ugly sores covered his scalp, and an intravenous line ran into the leathery skin of his forearm. "He is suffering from malnutrition and sleeping sickness," the Guinean nurse softly explained to me in French. Lying in the next bed was a young girl, maybe seven years old, also in deep coma. "This one has malarial meningitis," the nurse continued. I looked around at each bed and was overwhelmed by what I saw. But I hadn't noticed, when I had entered, the first bed to the right of the door. Lost in the vastness of the bed was a minute, shriveled baby, her matchstick arms and legs held loosely by webbed straps tied to the metal bed frame. She, too, had an intravenous line placed in her skinny wrist, and a stomach tube for force-feeding of milk was taped to the side of her mouth. "And this little one?" I asked the nurse. "She is two months premature," the nurse replied. There were no incubators in this hospital—no way to keep the infant at a constant warm temperature during the cool nights—nor were there screens on the windows or netting over the bed to protect her from the swarms of mosquitoes that would descend once the sun went down. As I watched, the baby began to quiver and writhe, her whole body racked by convulsions, and she let out a long, croaking screech, like the mewing of an old, dying cat. I bent down to stroke the side of her little face and shushed her to be still. Bit by bit, she relaxed until she fell back into a deep, silent sleep.

As I left the ward, after saying good-bye to the kind nurse, a painful lump rose in my throat, and I found I could utter no words. As Pam and I walked back to the car, I steeled myself not to shed a tear or show in any way the devastating effect these scenes had had upon me. I couldn't help thinking that somewhere, someone is doing research on malaria, sleeping sickness, and the other diseases that I had seen in that one ward, and I knew whoever it was would be using animals as experimental models.

I'm not sure why, but as Marie-Paule and I drove along the coast on our return to Black River, I began to think of my mother, who had died at the young age of fifty. What would she think of me? Would she understand why I was in research, she who had given me my love of animals and had tried to teach me how to be gentle and patient? She had been so proud of my becoming a vet, not because of any social status it would give me—although the vet in Ireland carries a great deal of that, and anyone who can shoe a horse or age him by his teeth is sure of a few pints of Guinness as recompense—but more out of her gentle concern for animals, or "crathers," as she called them. Would she have thought me cruel or uncaring because I now used animals in research?

I remembered phoning from Scotland her last Christmas Eve, barely two months after I had become a vet. I had just finished a visit to a farm to treat a cow for milk fever, and I was standing in a kiosk on the top of a windswept hill overlooking the Irish Sea, just outside the tiny village of Ballantrae, in Galloway. As I looked out through the glass side of the telephone booth, I could barely make out a thin strip of dark coastline on the horizon. "I can see Ireland as I'm talking to you, Mam," I said, and before I knew it, she was crying. We were only 450 miles apart, she at home in southern England, but in those days, with the difficulty of travel, it might as well have been ten times that distance. She had had a lonely and troubled life, she and my father not getting on well together since I was a very young child. It's strange, when I think back on it. My mother was a Catholic from the south of Ireland, and my father was a Protestant from the north—my grandfather even played the fife in the Orange parades every July 12, to mark King Billy's 1690 defeat of the Catholic forces at the Battle of the Boyne. And although they fought about nearly everything on the face of the earth, their arguments were never centered on religion or nationalism.

As I stood talking to her on the telephone, I tried to cheer my mother up by recounting an incident that had happened to me just a day or two before. As I had been driving home after a day of farm and house calls, a tiny kitten ran out from the side of the road under a car some distance ahead of me. The driver continued on his way, unaware that the kitten had been struck and now lay motionless in the middle of the road. I stopped the car and carefully picked her up. She was unconscious, and her chin was split so that I could wiggle the bones of each side of her jaw. The skin of her chin had been stripped off her jawbones and was hanging like a pouch at her throat. Unable to locate the owners of the kitten, I took her home, treated her with subcutaneous injections of fluid, tied the sides of her jaw together with silver wire threaded between her teeth, and stitched her chin back in place as best I could. Her prognosis seemed good, and the story seemed to lift my mother's spirits.

Not three weeks later, while I was on a farm calving a cow, the farmer's wife rushed into the byre to tell me that she had received a telephone call to say that my mother had taken seriously ill, was already in hospital, and that I should go to her as fast as possible. I finished the calving and drove straight to the airport in Glasgow, fifty miles away. I took the first plane to the south coast of England, not stopping to change my clothes or even my Wellington boots, which were covered in cow manure.

Within eight hours, I was at my mother's bedside. She was as white as a sheet, and so very, very tired. Unbeknownst to me at the time, X rays had revealed a dissecting aneurysm of her aorta. "How is the little kitten?" she asked me in her soft Irish brogue. "She's doing fine," I said. My mother looked up into my eyes, and with a faint smile she replied, "That's good!" and then she died.

As a boy, I would look up at my mother as she stood at the kitchen sink each evening. She would be peeling the potatoes with an old horn-handled knife, worn to a crescent from constant sharpening. As silent tears trickled down her cheeks, she would softly sing in her out-of-tune voice, songs that recalled the leaving of Ireland, and the potato famine, and the Troubles. On reflection, I was now sure that she too must have been constantly asking herself about the meaning of life.

It was almost dark by the time Marie-Paule and I made it back to the villa. We had our evening meal by candlelight, our last dinner in Jamaica; in the morning we would return to the States, back to reality. After supper, we sat out on the verandah, Marie-Paule to have a glass of wine, me to enjoy a last cool Red Stripe beer. I sat silent, trying to sort out my jumbled thoughts and wondering how they might be connected to the sadness I felt at the day's events.

I could hear, from my spot on the verandah, the coarse calls of the egrets as they returned to their roosts. I could make out the distant dark silhouette of the sixty-five-year-old Ambrose as he gently led his goats back from their daytime grazing. Their newborn kids bleated in panic as they chased their mothers in gamboling stride, trying desperately to keep up. Ambrose had watched over us from his tiny boat as Marie-Paule and I had explored the coral reefs far out to sea, had lifted us out of the water with the amazing strength of one wrist, had taught me how to extract the succulent juices from the sugarcane with a machete. This was the timeless rhythm of Jamaica. The sun would rise again tomorrow, and life would begin anew.

I couldn't help thinking of all the animals I had known over the years—six or seven hundred chimpanzees, the many hundreds of rhesus and Java monkeys, scores of baboons, and all the other animals that I had used in research in the endless search to help find cures for human diseases. I thought, too, of the haunting mistakes I had made as a vet over those years, and the elation of success after struggling through the night alongside the technicians to pull an animal through what I was sure was a terminal crisis. I thought, too, of the cowardly part I had played all those years ago in the deaths of Pops and Nico, the only chimps of the fourteen on LEMSIP's "surplus" list to be assigned to a terminal study. Pops, your regular, squat, nondescript chimp (not that that made any difference), and Nico, an unusually handsome and magnificent animal. Then my thoughts turned to Spike. Full-grown now, weighing about 130 pounds, he was all muscle and brawn. Yet Spike's only claim to fame had been his biting off the tip of Jane Goodall's thumb during one of her visits to the lab. Someone once described him to me as a little thick, rather slow to learn. "Have you noticed?" she asked. "Not only does his lower lip hang down all the time, but he constantly has boogers up his nose." For some reason Spike had reminded her, this keen observer of chimpanzee character, of a truck driver or an Irish boxer. Even I had to admit that he wasn't one of your rocket scientists among chimps, but he would always be my little Spike.

I had managed to get Spike, and eight other chimps, all veterans of research, into retirement recently at the Wildlife Waystation, a refuge for displaced and needy animals in southern California. Founded and run by Martine Colette, a small, tough French-born dynamo, the Waystation is nestled in the rolling hills and canyons of the Angeles National Forest, northeast of Los Angeles. Living by the motto of the Waystation, "Deeds, not words," Martine came through for me in my hour of need, after long months of desperate searching across the country, and so many empty promises of help from others.

Thanks to pressures from the animal rights movement and a growing awareness of public opinion, those scientists who, in the early 1980s, were advocating euthanasia of "surplus" chimpanzees as a way of avoiding the enormous financial burden of maintaining them, began to talk about setting up retirement facilities for the animals' lifelong care. But it has been largely talk, and forming committees and subcommittees to oversee the retirement facilities that, after twelve years, still don't exist.

Nine chimps was an insignificant number compared to the couple of hundred that had already earned their retirement, but at least it was a start. I traveled out to California to spend a few days with the chimps to help them settle in to their new life. When the time came to leave, I felt sad, yet elated. At last there would be no more sticking and poking with needles, no more living alone in a cage in a dreary lab. Spike and the others could look forward to a lifetime of freedom and peace.

What was so special about Spike? Why did I choose him to be among the small group of chimpanzees that I was trying to quietly send into retirement? After all, Spike was young, and many of the other chimpanzees going into retirement had given far more years of their lives to research and were far more deserving of special consideration. Rufe, for instance, was the elder statesman of the group at thirty-nine years of age. He had come close to dying from heart disease only four years earlier. One of the most gentle and intelligent of chimpanzees, Rufe had participated in the aerospace program in his earlier days at the Holloman Air Force Base in New Mexico, before arriving at LEMSIP in the late 1960s to undergo years of research into viral hepatitis.

Booee was the only famous member of the group going into retirement. In fact, he was the only famous chimp we had at LEMSIP. Booee had been part of the chimpanzee American Sign Language program at the University of Oklahoma, in Norman. The Institute of Primate Studies there ran into financial difficulties in the early 1980s, and the director, Dr. Bill Lemmon, was forced to disband the colony. Many of the animals arrived at LEMSIP to continue their research careers, but this time in biomedical research.

Booee once accused me of lying, in his nonstop signing. I had entered the room where he and nine other animals were caged. Booee asked me for a treat, quickly running his index finger across his scalp—the sign for his name—then hitting his forehead to indicate "wants," and finally touching the side of his cheek to say "treat." I did have sweets in my pocket, pink-and-white striped peppermints that I purloin in great numbers from restaurant counters specially for the chimps. However, I didn't have enough to give one to every animal in the room, and I knew Booee couldn't see the candies anyhow. So I lied, signing to him that, no, I didn't have any sweets. He asked me again for one. I lied a second time. With lighting speed, Booee repeated twice in quick succession "Booee see [as he pointed emphatically to his eye] treat in pocket."

Booee broke many hearts when he appeared on television in early 1995. Portrayed as the animal who had been marooned in research, there was an outpouring of sympathy for him, from the general public, the Great Ape Project (founded by Peter Singer, the Australian animal rights philosopher), many of the animal rights groups, and the Jane Goodall Institute. Yet no one showed the least bit of interest in the other chimps; they just weren't famous and newsworthy enough.

Two of the nine retiring chimpanzees, Chas and Seetee, had been used in the AIDS program at LEMSIP, although they hadn't been infected with live virus. Lindsey, Ray, and Bold had participated in many studies and had even had their spleens removed in an unsuccessful program to develop vaccines against malaria. In contrast, Spike had been used in only two major projects: the first a safety test for a hepatitis vaccine, the second a safety test of a blood clotting factor prepared from pooled human blood, to ensure that it contained no hidden viruses before it was used in human trials.

Yet when does even a scientist, seeing no alternative to using animals in research, say "This animal has paid his dues"? Is it after a certain amount of time, or a prescribed number of research projects? Some chimpanzees are robust and seem to be able to put up with any amount of use, while others can be psychologically demolished by even a single project.

WITH SPIKE, I FELT an inextricable bond, as though we had drawn up an unspoken agreement. After all, I had brought him into the world and given him his first breath of life. We had shared food. Like a mother must feel for her children, I felt an inescapable responsibility towards Spike. But, at the same time, I had to admit that neither study Spike had been put on was particularly taxing, in a physical sense, especially to a big, strong animal like him, but a virus, unknown to medical science at the time, may well have been lurking in the blood clotting preparation he had received in the second study.

Without my realizing it, Spike had a time bomb ticking away inside him, ready to explode if even the mildest illness struck him. Within barely a month of arriving at the Waystation, Spike became seriously ill. I flew out again to see him.

Dehydrated and delirious, Spike didn't even recognize me. He had lost a lot of weight in only forty-eight hours, and he was deeply jaundiced. I and the two resident veterinarians at the Waystation, Becky Yates and Silvio Santinelli, immediately set about taking blood samples from Spike in an attempt to diagnose what was ailing him. In the meantime, we hooked him up to an intravenous line to infuse him with much-needed fluids and electrolytes. Spike's liver was failing, that was patently obvious, but without the blood test results we couldn't determine whether he was suffering from primary liver failure or whether his liver condition was secondary to some other disease.

In the meantime, Spike needed plasma to relieve the burden on his liver, and once again I found myself calling for Dr. Socha's help to send, as fast as possible, the best matched plasma he could find from the other chimps at LEMSIP. Becky, Silvio, and I worked feverishly on Spike over the following days, and by the third transfusion of plasma, he began to make a spectacular recovery. His appetite returned—in fact he was eating like a horse, as though he was making up for lost time—and he was back to his playful, rambunctious self. Greatly encouraged by his progress and anxious about the press of work back at LEMSIP, I saw no need to stay any longer. I knew I could leave Spike in Becky and Silvio's capable hands.

I remember so well that last morning when I dashed in to see Spike once more before leaving the Waystation to catch my early flight back to New York. In the subdued light of predawn, I quietly approached Spike's cage, not wanting to disturb him too much or awaken the other chimps, just to tell him a quiet good-bye, and to say that I would be back again to see him soon. He heard my coming and stirred himself out of sleep in his deep-strawed nest to lumber over towards me. Spike still looked very gaunt, his gait stiff and awkward, but the brightness had returned to his eyes and his spirit was high. With his fingers stretched through the bars of the cage, he gently grasped the back of my head in both hands and pressed my face to the bars so that he could give me a great, open-mouthed, huffing kiss. Little could I have realized that this would be the last time I would see him. Not four days later, just two days short of his sixteenth birthday, Spike died in his sleep, his liver shot to pieces, and a light went out in my soul.

Doug Cohn, my young veterinary colleague at the lab, tried to comfort me. "You know, Jim," he said, "Spike saw the blue skies and heard the birds sing. He made it to the Promised Land."

Perhaps that's what life is all about, I said to myself, and my thoughts returned to Molly. I could see her in my mind's eye, bounding through the forest, her head and neck outstretched, her legs extended, gliding through the air with the grace of a deer, consumed by the joy of living, this little thing that had struggled so hard to survive, and I thought, Molly, this little one-eyed Jamaican bush dog, has also made it to the Promised Land.

Afterword

■ ■ ■

MUCH HAS CHANGED SINCE Jim Mahoney finished writing this book. His laboratory has closed and the chimps and monkeys are gone, but Molly's still going strong.

On December 31, 1997, LEMSIP closed its doors, thus ending the existence of a laboratory that PETA founder Alex Pacheco said was "different from all the other laboratories." The Coulston Foundation, a New Mexico-based toxicological laboratory, did not take over after all but did get almost half of the LEMSIP's two hundred chimpanzees gratis. Mahoney, often ignoring the wishes of his superiors, sought and found retirement for the rest. On one eventful night, Mahoney arranged to smuggle thirty-two young chimps—the most valuable because they were yet to be used as research subjects—out of the laboratory in two trucks driven by men he did not know. The plan was to spirit them away to a refuge in California. On the way, U.S. Fish and Wildlife Service agents, acting on a tip that animals were being smuggled for illegal sale, stopped the trucks and arrested one of the drivers on drug and weapons violations. Even so, the chimps made it safely to California.

"My school didn't trust me, and the animal rights groups thought I was working against the best interests of the animals," Mahoney says. "I was accused of being traitorous, disingenuous, and deceitful, and of working with criminal elements."

Mahoney doesn't understand why he wasn't fired after this incident. His clandestine activities required that he make painful choices about which chimps would remain caged research subjects and which would go to a life of freedom in retirement and which would remain caged research subjects for the rest of their lives. "I hated myself for it and could hardly bring myself to look the animals in the eye afterwards," he says now.

For the moment, Mahoney remains a member of the faculty of the New York University Medical School but will unlikely pursue another position in medical research after all that has happened. He would like to continue working with the animals he helped to place in sanctuaries and to combine this with his interests in chimpanzee rehabilitation in Africa. "There's a lot to be done," he says.

Molly is an elegant dog now, with a beautiful, sleek outline. She runs free, as she would have had she stayed and survived in Jamaica. When Mahoney goes to sit in his African chair with a beer to ruminate after he arrives home from work, she's always waiting there to sit with him. She has a boyfriend named Whiskey who lives on a nearby hill.

the Editors

In Appreciation

■ ■ ■

THE AUTHOR AND ALL the employees of the former LEMSIP gratefully extend their thanks to the organizations listed below, who came forth in the animal's hour of need and offered them the prospect of dignity and freedom they so much deserved after their years of service to mankind. Over one hundred monkeys (rhesus and Java macaques, baboons, cotton-eared marmosets and red-bellied tamarins) and almost ninety chimpanzees finally reached sanctuary.

BLACK BEAUTY RANCH
Fund for Animals
Murchison, Texas

LA FONDATION FAUNA
Carignon, Quebec
Canada

PACIFIC PRIMATE SANCTUARY
Maui, Hawaii

Primarily Primates Inc.
San Antonio, Texas

Primate Rescue Center
Nicholasville, Kentucky

The Sanctuary for Animals
West Town, New York

Trevor Zoo
Millbrook School
Millbrook, New York

Wild Adventure
Valdosta, Georgia

Wild Animal Orphanage
San Antonio, Texas

Wildlife Rescue
Boerne, Texas

The Wildlife Waystation
Angeles National Forest
California